T0134904

Sustainable Civil Infrastructures

Editor-in-Chief

Hany Farouk Shehata, SSIGE, Soil-Interaction Group in Egypt SSIGE, Cairo, Egypt

Advisory Editors

Khalid M. ElZahaby, Housing and Building National Research Center, Giza, Egypt
Dar Hao Chen, Austin, TX, USA

Sustainable Infrastructure impacts our well-being and day-to-day lives. The infrastructures we are building today will shape our lives tomorrow. The complex and diverse nature of the impacts due to weather extremes on transportation and civil infrastructures can be seen in our roadways, bridges, and buildings. Extreme summer temperatures, droughts, flash floods, and rising numbers of freeze-thaw cycles pose challenges for civil infrastructure and can endanger public safety. We constantly hear how civil infrastructures need constant attention, preservation, and upgrading. Such improvements and developments would obviously benefit from our desired book series that provide sustainable engineering materials and designs. The economic impact is huge and much research has been conducted worldwide. The future holds many opportunities, not only for researchers in a given country, but also for the worldwide field engineers who apply and implement these technologies. We believe that no approach can succeed if it does not unite the efforts of various engineering disciplines from all over the world under one umbrella to offer a beacon of modern solutions to the global infrastructure. Experts from the various engineering disciplines around the globe will participate in this series, including: Geotechnical, Geological, Geoscience, Petroleum, Structural, Transportation, Bridge, Infrastructure, Energy, Architectural, Chemical and Materials, and other related Engineering disciplines.

More information about this series at http://www.springer.com/series/15140

Laureano Hoyos · Hany Shehata
Editors

Advancements in Unsaturated Soil Mechanics

Proceedings of the 3rd GeoMEast
International Congress and Exhibition, Egypt
2019 on Sustainable Civil Infrastructures –
The Official International Congress
of the Soil-Structure Interaction Group
in Egypt (SSIGE)

 Springer

Editors
Laureano Hoyos
The University of Texas
at Arlington
Texas, TX, USA

Hany Shehata
Soil-Structure Interaction Group
in Egypt (SSIGE)
Cairo, Egypt

ISSN 2366-3405   ISSN 2366-3413   (electronic)
Sustainable Civil Infrastructures
ISBN 978-3-030-34205-0   ISBN 978-3-030-34206-7   (eBook)
https://doi.org/10.1007/978-3-030-34206-7

This Springer imprint is published by the registered company Springer Nature Switzerland AG
The registered company address is: Gewerbestrasse 11, 6330 Cham, Switzerland

Contents

About the Editors

Laureano Hoyos Ph.D., P.E., M.ASCE

Professional Preparation:

Post-Doctoral Research Fellow, Louisiana State University, 1999

Post-Doctoral Research Fellow, Georgia Institute of Technology, 1999

Ph.D. Civil and Environmental Engineering, Georgia Institute of Technology, 1998

M.Sc. Civil and Environmental Engineering, Georgia Institute of Technology, 1996

M.Sc. Geotechnical Engineering, University of Puerto Rico-RUM, Mayaguez, PR, 1993

M.Sc. Highway Engineering, Universidad del Cauca, Popayán, Colombia, 1991

B.Sc. Civil Engineering, Universidad de la Costa, Barranquilla, Colombia, 1988

Appointments:

- Chair, Unsaturated Soils Committee, Geo-Institute of ASCE
- Professor, Department of Civil Engineering, UTA, Texas, 2014–Current
- Associate Professor, Department of Civil Engineering, UTA, Texas, 2005–2014
- Assistant Professor, Department of Civil Engineering, UTA, Texas, 2000–2005

Honors:
- Lockheed Martin Aeronautics Excellence in Teaching Award, College of Engineering, University of Texas at Arlington, 2014
- Research Excellence Award, Office of the Provost, University of Texas at Arlington, 2006, 2007, 2008, and 2009
- Outstanding Civil Engineering Instructor Award, Department of Civil Engineering, University of Texas at Arlington, 2005
- Outstanding Early Career Faculty Award, College of Engineering, University of Texas at Arlington, 2003

Hany Shehata is Founder and CEO of the Soil-Structure Interaction Group in Egypt "SSIGE." He is Partner and Vice-President of EHE-Consulting Group in the Middle East and Managing Editor of the "Innovative Infrastructure Solutions" journal, published by Springer. He worked in the field of civil engineering early, while studying, with Bechtel Egypt Contracting & PM Company, LLC. His professional experience includes working in culverts, small tunnels, pipe installation, earth reinforcement, soil stabilization, and small bridges. He also has been involved in teaching, research, and consulting. His areas of specialization include static and dynamic soil–structure interactions involving buildings, roads, water structures, retaining walls, earth reinforcement, and bridges, as well as, different disciplines of project management and contract administration. He is Author of an Arabic practical book titled "Practical Solutions for Different Geotechnical Works: The Practical Engineers' Guidelines." He is currently working on a new book titled "Soil-Foundation-Superstructure Interaction: Structural Integration." He is Contributor of more than 50 publications in national and international conferences and journals. He served as Co-chair of the GeoChina 2016 International Conference in Shandong, China. He serves also as Co-chair and Secretary-General of the GeoMEast 2017 International Conference in Sharm El-Sheikh, Egypt. 2016 Outstanding reviewer of the ASCE as selected by the Editorial Board of International Journal of Geomechanics.

Characterization of the Mechanical Properties of Sensitive Clay by Means of Indentation Tests

Vincenzo Silvestri[1($\boxtimes$)] and Claudette Tabib[2]

[1] École Polytechnique, Montreal, Canada
vincenzo.silvestri@polymtl.ca
[2] Montreal, Canada

Abstract. Instrumented indentation tests were carried out on specimens of undisturbed clay, using various indenters at a constant rate of penetration of 0.5 mm/min. These tests were performed with 20-mm in diameter conical indenters with apical angles of 60°, 80°, 97°, and 144.3°, as well as with a Vickers indenter of semi-apical angle of 68.7°. Such tests allowed the determination of both the undrained shear strength and the deformation modulus of the clay. The undrained shear strength was found by assuming that a fully plastic stress field resulted from indenter penetration and by using slip-line theory. The deformation modulus was computed by assuming that the clay behaved elastically during initial penetration. The theory proposed by Love (1939) and later extended by Sneddon (1948, 1965) was used to obtain values of Young's modulus. The expanding cavity model or ECM proposed by Johnson (1970, 1985) was also employed for the interpretation of test results obtained with both sharp and blunt indenters.

One of the most significant findings of the present study is that the structure of the undisturbed clay suffers severe and progressive breakdown with increasing penetration during indentation, which results in a dramatic decrease of the strength and deformation parameters of the clay. Comparison between the undrained shear strength deduced from the quasi-static indentation tests and the dynamic Swedish fall-cone tests, also indicates that S_u obtained from the latter tests are much higher than the corresponding data derived from the indentation tests. It is believed that the possible cause of the overestimation of the undrained shear strength deduced from the Swedish fall-cone tests is related to the very high strain rate experienced by the falling cone.

The paper also presents a brief review of the most pertinent theories that are used for the interpretation of indentation tests.

Keywords: Indentation tests · Conical and pyramidal indenters · Sensitive clay · Young's modulus · Undrained shear strength · Comparisons

1 Introduction

Different test methods are available for the determination of the mechanical properties of materials. The results of these methods are employed for the design of engineering structures and as a basis for comparison and selection of materials. For instance, uniaxial tension and compression tests are commonly used to determine mechanical

© Springer Nature Switzerland AG 2020
L. Hoyos and H. Shehata (Eds.): GeoMEast 2019, SUCI, pp. 1–18, 2020.
https://doi.org/10.1007/978-3-030-34206-7_1

properties of metals and alloys, like Young's modulus, yield strength, and strain-hardening characteristics. In geotechnical engineering, triaxial tests, unconfined compression tests, simple shear tests, and Swedish fall-cone tests fulfill a similar role in the determination of the short-term strength parameters of clays.

As an alternative and appropriate method to determine the mechanical properties of materials, indentation tests have been proposed by many researchers (Tabor 1951; Johnson 1970, 1985; Oliver and Pharr 1992; Fischer-Cripps 2002, 2007). The main advantages of indentation tests are the following: (a) They do not require a large quantity of material: a well-polished and flat surface is sufficient to generate a large set of data; and (b) The measurement is non-destructive in nature as the impression is generally confined to a small region of the sample. According to the load-penetration data recorded during indentation, the mechanical properties of the specimen including Young's modulus E, yield strength σ_y, hardness H, and viscoelastic deformation parameters may be determined (Tabor 1951; Johnson 1985; Oliver and Pharr 1992; Fischer-Cripps 2007). The International Organization for Standardization (ISO) and the American Society for Testing and Materials (ASTM) issued standards which cover depth-sensing indentation testing for determining hardness and other material parameters (ISO 2002, 2007; ASTM 2003, 2007).

It is generally agreed that hardness is a measure of a material's resistance to permanent penetration by another harder material (Fischer-Cripps 2002). Of particular interest in indentation testing is the area of the contact zone found from the dimensions of the contact perimeter. The mean contact pressure p_m, as given by the indenter load P divided by the area of contact A, is a useful normalizing parameter; it has the additional advantage of having actual physical meaning (Fischer-Cripps 2002). The value of the mean contact pressure at which there is no additional increase in load is related to the hardness (Tabor 1951). In addition, the mean contact pressure is taken equal to the hardness H for indentation test methods that employ the projected area of contact A_p at this limiting condition.

Theoretical analyses of indentation problems have received attention from a number of investigators, since Prandtl (1920) developed a slip line solution for a two-dimensional punch on a semi-infinite plastic medium. A detailed analysis of wedge indentation was given by Hill et al. (1947) and Hill (1950), and Shield (1955) obtained a similar solution for axisymmetric indentation by a flat circular punch. These solutions apply to rigid plastic materials and are based on the slip line theory. Later, Lockett (1963) performed a numerical analysis of conical indentation based on the slip line theory and obtained solutions for apical angles $2\alpha \geq 105°$. Lockett's solution was extended to other conical indenters by Houlsby and Wroth (1982) and Chitkara and Butt (1992).

Another line of approach to understanding indentation is semi-empirical in nature. Tabor (1951) presented a correlation between the mean contact pressure p_m and the yield stress σ_y. The correlation is expressed as:

$$p_m = H = C \, \sigma_y \tag{1}$$

where C is a constraint factor, the value of which depends on the type of specimen and the geometry of the indenter. For example, C = 3 for materials with a large ratio E/σ_y

(i.e., metals) and C = 1 to 1.5 for low values of E/σ_y (i.e., glasses and ceramics). To describe the correlation in Eq. 1, Johnson (1970) adopted the model of Marsh (1964) who, following Bishop et al. (1945) and Samuels and Mulhearn (1957), considered the plastic deformed zone beneath a blunt indenter as the expansion of a spherical cavity in an elastic-plastic solid under an internal hydrostatic pressure. Johnson (1970) included the effect of the indenter geometry.

In spite of these analyses, detailed solutions of indentation problems for realistic stress-strain relationships and indenter shapes are still lacking because of the very complex stress and strain fields produced by indentations. To resolve some of the uncertainties, indentation problems have been also approached using the finite element method of analysis in which indentations are simulated by means of rigid or elastic indenters on elastic, elastic-plastic, and viscoelastic materials (Barquins and Maugis 1982; Bhattacharya and Nix 1988; Giannakopoulos et al. 1994; Cheng and Li 2000; Riccardi and Montanari 2004; Guha et al. 2014; Hu et al. 2015).

While a large number of analytic and numerical studies on indentation have been conducted, commensurate experimental investigations, whose primary aim is to understand the elastic-plastic flow field beneath the indenter, are less common. Keeping this in view, an experimental study was undertaken on the influence of the cone indenter apical angle on the determination of the shear strength and Young's modulus of undisturbed overconsolidated sensitive clay of eastern Canada. Conical indenters with apical angles of 60°, 80°, 97°, and 144.3°, as well as a Vickers indenter (a square pyramid with opposite faces at an angle of 134.6° and edges at 147°) were used in the test program.

The present paper is organized in the following manner. In the next section, pertinent experiments, theoretical analyses, and simulations that have been conducted previously are reviewed with the objective of putting the current work in proper perspective. Thereafter, the section on the test program describes the clay and the experiments conducted in this study, whereas the section on the analysis presents results, discussions, and comparisons. The paper closes with conclusions and some brief remarks about future directions in research in this area.

2 Background

2.1 Conventional Analysis of Indentation Tests

In a typical indentation test, load P and depth of penetration h are recorded as load is applied from zero to some maximum load P_{max} and then from maximum load back to zero. If plastic deformation occurs, then there is a residual impression left in the surface of the specimen. When load is removed from the indenter, the material attempts to regain its original shape, but is prevented from doing so because of plastic deformation. However, there is some degree of recovery due to the relaxation of the elastic strains within the material (Fischer-Cripps 2002). The form of the compliance curves, that is, load versus depth of penetration curves, for the most common types of indenter (i.e., spherical, pyramidal, and conical) are very similar and a typical curve is shown in Fig. 1. In this figure (see also Fig. 2), h_r is the depth of the residual impression, h_{max} is the depth from the original surface at maximum load, h_e is the elastic recovery during unloading, h_a is the distance from the edge of the circle of contact, as shown in Fig. 2, $h_p = h_{max} - h_a$, and dP/dh which is the slope of the initial portion of the unloading curve gives an estimate of the elastic modulus of the material.

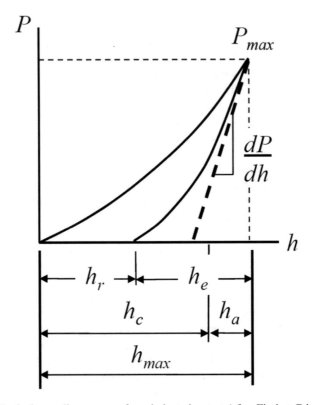

Fig. 1. Typical compliance curve from indentation test (after Fischer-Cripps 2007)

Indentation testing in many materials results in both elastic and plastic deformation of the indented specimen. In brittle materials, plastic deformation most commonly occurs with sharp indenters such as conical and pyramidal indenters. In ductile materials, plasticity may be readily induced with a blunt indenter such as a sphere or cylindrical flat-headed punch. Indentation tests are routinely used in the measurement of hardness, but conical and pyramidal indenters may be used to investigate other mechanical properties such as strength, toughness, and internal residual stresses (Yu and Blanchard 1996; Zeng and Chiu 2001; Fischer-Cripps 2002). Indentation tests involving spherical or conical indenters, first used as the basis for theories of hardness, enabled various criteria to be established. The most well-known criterion is that of Hertz, who postulated that an absolute value of hardness was the least value of pressure beneath a spherical indenter necessary to produce a permanent deformation at the center of the area of contact.

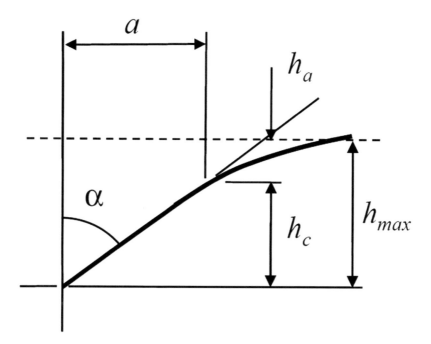

Fig. 2. Geometry of conical indentation (after Fischer-Cripps 2007)

The mean contact pressure, and, hence, the indentation hardness, for an impression made with a spherical indenter is given by:

$$p_m = H = 4P/\pi d^2 = P/\pi a^2 \tag{2}$$

where a is the radius and d is the diameter of the contact circle (the latter is assumed to be equal to the diameter of the residual impression left in the surface of the specimen upon removal of the load). The mean contact pressure determined in this way is often called the Meyer hardness. Meyer (1908) found that there was an empirical relationship between the diameter of the residual impression and the applied load, and this is known as Meyer's law:

$$P = k \, d^n \tag{3}$$

where k and n are constants for the specimen material. It has been shown that the value of n is insensitive to the radius of the spherical indenter and is related to the strain-hardening exponent x of the material according to (Fischer-Cripps 2002):

$$n = x + 2 \tag{4}$$

Values of n were found to range between 2 and 2.5, the higher value applying to annealed materials, while the lower value applying to work-hardened materials (i.e., low values of x). The strain-hardening exponent x plays an important role in the constitutive relationships of elastic-plastic materials, as the mechanical properties of specimens can be approximated by a uniaxial stress-strain response given by:

$$\sigma = E \in, \ \sigma \le \sigma_y/E \tag{5a}$$

and

$$\sigma = b \ \epsilon^x, \ \sigma \ge \sigma_y/E \tag{5b}$$

where σ is the applied stress, $\in$ is the resulting strain, and b is defined as

$$b = \sigma_y (E/\sigma_y)^x \tag{6}$$

The material is elastic-plastic for $x = 0$.

Remark 1
For $x = 0$ and $n = 2$, Eq. 3 which reduces to $P = k \ d^2$ is often called Fick's law. In addition, for a conical indenter of apical angle 2α, substitution of the radius of contact a for the depth of penetration h_p (Fig. 2) results in $P = k \ h^2 \tan^2\alpha$ for Fick's law or more generally, $P = k^* \ h^2$, where $k^* = k \tan^2\alpha$.

Remark 2
Indenters can be classified into two categories- sharp or blunt. The criteria upon which a particular indenter is classified, however, are subjective. For example, some authors classify sharp indenters as those resulting in permanent deformation in the specimen upon removal of the load. However, others prefer to classify a conical or pyramidal indenter having an equivalent cone semi-apical angle $\alpha \ge 70°$ as being blunt. For these, the response of the material follows that predicted by the expanding cavity model proposed by Johnson (1970) or the elastic constraint model of Tabor (1951), depending on the type of specimen material and magnitude of the load. For sharp indenters, it is generally observed that plastic flow occurs according to the slip line theory and the specimen behaves as a rigid plastic material.

2.2 Elastic Approach

The classical approach to finding the stresses and displacements in an elastic half-space produced by surface tractions is due to Boussinesq (1985) and Cerruti (1882) who made use of the theory of potential. Partial results based upon such theory were obtained by Love (1939) for the case of an elastic half-space indented by a rigid cone. Because Boussinesq's method does not lend itself to practical analysis, Sneddon (1948) used the integral transform technique to obtain the same result and to evaluate all the components of stress within the solid. In addition, Sneddon (1965) derived a relationship between the load P and the depth of penetration h in the axisymmetric

Boussinesq problem for a rigid punch of arbitrary shape. For a conical indenter of apical angle 2α, Sneddon's relationship reduces to (see also Fig. 2):

$$P = \pi a^2 E \cot \alpha / 2 \left(1 - v^2\right) \tag{7}$$

where E and v are respectively Young's modulus and Poisson's ratio of the specimen material, and a is the radius of the circle of contact. The depth profile of the deformed surface within the area of contact is

$$h(r) = (\pi/2 - r/a) \, a \cot \alpha \tag{8}$$

At the circle of contact, the quantity $a \cot \alpha$ represents the depth of penetration h_p. Substitution of Eq. 8 for $r = 0$ in Eq. 7 yields

$$P = 2 E h^2 \tan \alpha / \pi \left(1 - v^2\right) \tag{9}$$

where h is the depth of penetration of the apex of the cone beneath the original surface of the indented specimen. In addition, the distribution of the pressure $p(r)$ on the face of the conical indenter is given by

$$p(r) = \left(P/\pi a^2\right) \cosh^{-1}(r/a), \, r \leq a \tag{10}$$

which shows that there is a singularity at $r = 0$ and that $p(r) = 0$ at $r = a$.

Comparison between Meyer's law (Eq. 3) for $n = 2$ and Sneddon's expression (Eq. 9) shows that if the material behaved elastically, then the constant k becomes equal to $2 E \tan \alpha/\pi (1 - v^2)$ because the diameter of the circle of contact is equal to $2 h \tan \alpha$ for a conical punch. In addition, as the initial portion of the unloading curve of slope dP/dh is often considered to represent elastic response (Fischer-Cripps 2002), application of Eq. 7 allows determination of Young's modulus E,

$$E = P_{max} \pi \left(1 - v^2\right) / 2 h_e^2 \tan \alpha \tag{11}$$

since $dP/dh = P_{max}/h_e$. However, because the initial portions of the compliance curves obtained for the clay in the present investigation were either vertical or characterized by negative slopes (i.e., increasing penetration with decreasing load), Eq. 11 could not be used to determine values of Young's modulus. As a consequence, the elastic modulus was determined in a different way, as discussed later in the paper.

Remark 3

In indentation testing, pyramidal indenters, like the Vickers pyramid, are generally treated as conical indenters with a cone angle that provides the same area to depth relationship as the actual indenter, despite the availability of contact solutions for pyramidal indenter problems (Fischer-Cripps 2002). This allows the use of axial-symmetric elastic solutions, Eqs. 7 to 10, to be applied to contacts involving non-axial-symmetric indenters. Thus, as the relationship between the projected area of contact A_p and the penetration h_p in Fig. 2 is given by $A_p = 4 h_p^2 \tan^2\theta$ for a Vickers pyramid with a face angle θ, whereas the relationship is $A_p = \pi h_p^2 \tan^2\alpha$ for a conical indenter, then the Vickers pyramid may be treated as a conical indenter with an effective cone angle given by $\tan^2\alpha = 4 \tan^2\theta/\pi$, resulting, for example, in $\alpha = 69.7°$ for $\theta = 67.8°$.

2.3   Rigid-Plastic and Elastic-Plastic Approach

Hill et al. (1947) were the first to examine the plastic flow response of metals indented with different wedge indenters having α varying between 7° and 30°, and concluded that the plastic flow beneath a sharp indenter is cutting in nature. Consequently, Hill (1950) proposed the slip line theory, in which the material beneath the indenter is displaced laterally and upwards from the sides of the wedge. The theory was validated by Dugdale (1953) who performed wedge indentations on cold-worked metals. Atkins and Tabor (1965) systematically studied the indentation response of various metals with conical indenters having α between 30° and 75°, and found that there was a change in plastic flow mechanism from cutting to compressive type at α equal to about 52.5°. It should be also recalled that Shield (1955) obtained a solution for a flat cylindrical punch and Lockett (1963) extended Shield's approach to conical indenters, but was unable to obtain a solution for $\alpha \leq 52.5°$. Later, Houlsby and Wroth (1982) and Chitkara and Butt (1992) obtained additional solutions. Figure 3 which summarizes the above-mentioned solutions presents values of the so-called cone factor N_c as function of the apical angle 2α for smooth rigid cones. The mean contact pressure at yield p_m is directly related to the undrained shear strength through the factor N_c, that is,

$$p_m = N_c S_u \tag{12}$$

where $S_u = \sigma_y/2$ for Tresca criterion.

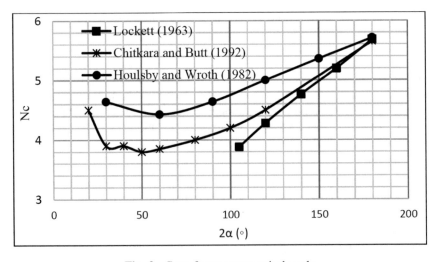

Fig. 3. Cone factor versus apical angle

Samuels and Mulhearn (1957) examined the plastic flow beneath spherical and Vickers indenters and noted that the material under the indenter flowed radially outwards and the elastic-plastic boundary appeared to be hemispherical in shape. These findings were rationalized by recourse to the spherical cavity model of Bishop et al. (1945) and Hill (1950). Later, Mulhearn (1959) conducted indentations experiments on work-hardened steel and concluded that plastic flow occurred via a cutting mechanism under a sharp indenter ($\alpha = 20°$), whereas it was compressive in nature for relatively blunt indenters such as the Vickers pyramid ($\theta = 68°$, $\alpha = 70.3°$).

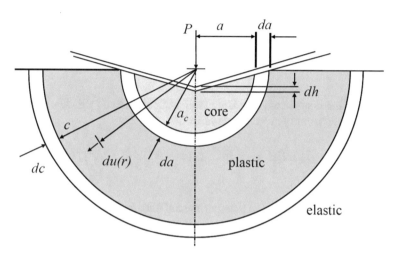

Fig. 4. Expanding cavity model (after Johnson 1970 and Fischer-Cripps 2007)

Dugdale (1954) carried out Vickers indentations on different metals and found that the spherical cavity model could not be used to represent the nature of deformation, because the mean contact pressure needed to expand the cavity was almost twice that required for a shallow penetration. The same observations were made by Marsh (1964) and Hirst and Howse (1969). This led them to suggest that the spherical cavity model had to be modified. On the basis of these results, Johnson (1970) proposed the expanding cavity model or ECM to describe the mean contact pressure for elastic-plastic solids. In this model, the contacting surface of the indenter is encased by a hydrostatic core of radius a_c which in turn is surrounded by a hemispherical plastic core of radius c, as shown in Fig. 4. An increment of penetration dh of the indenter results in an expansion of the core da and the volume displaced by the indenter is accommodated by the radial movement du(r) at the core boundary. This in turn causes the plastic zone to increase in radius by an amount dc. Using this result Johnson (1979) showed that the mean contact pressure p_m for elastic-plastic solids is:

$$p_m = \frac{4}{3}\sigma_y + \frac{2}{3}\sigma_y \ln\left\{\left[E\cot\alpha/\sigma_y + 4\left(1 - 2v\right)\right]/6\left(1 - v^2\right)\right\} \qquad (13)$$

and this leads, from Eq. 1, to the following expression for the constraint factor C:

$$C = \frac{4}{3} + \frac{2}{3}\ln\left\{\left[E\cot\alpha/\sigma_y + 4(1 - 2v)\right]/6(1 - v^2)\right\} \tag{14}$$

Although the ECM constitutes an idealization of the deformation mechanism that takes place beneath a blunt indenter, it nevertheless represents a class of popular models that have the capability to consider the effects of elastic and plastic response on the evolution of the deformation field during indentation (Wang et al. 2016).

In addition, Hainsworth et al. (1996) showed that the loading curve of a specimen could be described using a linear relationship between the load P and the square of the displacement h:

$$P = A^* h^2 \tag{15}$$

Superimposing the displacements arising from both elastic and plastic deformation, the constant of proportionality A* was found to be:

$$A^* = E\left\{[1/(\pi \tan^2\alpha)]\left[E/H(1 - v^2)\right]^{1/2} + (\pi - 2)\left[H(1 - v^2)/E\right]^{1/2}\right\}^{-2}/(1 - v^2) \tag{16}$$

This approach provided a good fit to experimental data for specimens with a wide range of modulus and hardness values.

It was also realized in the course of the present study that it is possible to determine the mechanical properties of interest in a "forward" direction rather than in the "reverse" direction, which involves the determination of shape of indentation test results for the deduction of material properties. In the "forward" direction approach, using material parameters (for instance, values of Young's modulus E and yield stress σ_y obtained from unconfined compression tests) as inputs, the expected load-displacement response can be predicted and compared with that obtained experimentally. Alternatively, the expected P vs h^2 relationship for a sharp indenter may be used to determine E and the hardness H by fitting this function to the experimental loading curve. In the present study both approaches were attempted as discussed in the next section.

3 Experimental Investigations

The experimental investigations reported in the present paper were carried out by Ewane (2018). A detailed description of the soil and test procedures may be also found in Ewane et al. (2018).

3.1 Clay

The soil used in this study is an overconsolidated clay of eastern Canada. Undisturbed blocks of clay, 0.3 m in width, were recovered at a depth of 3.5 m in a test excavation. The test site is located 40 km east of Montreal (Quebec), in the town of Beloeil.

The natural water content of the clay varies from 43.7 to 63%, the plastic limit from 20.4 to 33.6%, and the liquid limit from 48.4 to 68%. The clay is overconsolidated, with an overconsolidation ratio (OCR), which is the ratio between the preconsolidation pressure σ_p' and the in situ vertical effective stress σ_{vo}', varying between approximately 5.8 and 8.4. Values of undrained shear strength S_u were determined using miniature vane tests (VST), unconfined compression tests (UU), and Swedish fall-cone tests. Average values of S_u were found to be 33.5 kPa from VSTs, 56.5 kPa from UU tests, and 73.4 kPa from Swedish fall-cone tests. While the lowest value of 33.5 kPa is attributed to the development of cracks around the blades of the miniature vanes upon their insertion, the value of 73.4 kPa obtained from the fall-cone tests is extremely high for such medium overconsolidated clay. As the average preconsolidation pressure σ_p' of the clay is 143.2 kPa, the strength ratio S_u/σ_p' equals 0.233 for the VSTs, 0.396 for the UU tests, and 0.512 for the fall-cone tests. Because measurements have repeatedly shown that the strength ratio for this type of clay falls in the range of 0.25–0.30 (Leroueil et al. 1983), it is believed that the very high value of 0.512 is possibly the result of the very high penetration rate that occurs during the penetration of the falling cone. Typical unconfined compression test results are reported in Fig. 5. Values of Young's modulus E range between 15 and 20 MPa, whereas values of yield stress σ_y vary from 104 to 108 kPa. This figure also shows that the clay may be treated as an elastic perfectly plastic material with a strain-hardening exponent x = 0. Additional UU tests yielded an average value of 113.4 kPa for σ_y which corresponds to $S_u = \sigma_y/2$ (Tresca) or 56.7 kPa.

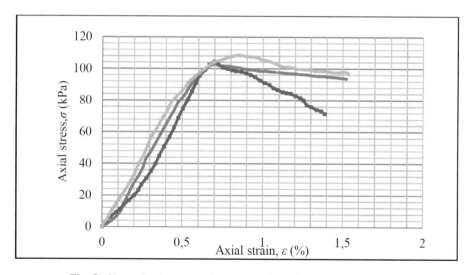

Fig. 5. Unconfined compression test results (after Ewane et al. 2018)

3.2   Indentation Tests

Tests were carried out using a computer-controlled universal testing machine. Indentation tests were performed with 20-mm diameter conical indenters with apical angles 2α of 60°, 80°, 97°, and 144.3°, as well as with a square-sided Vickers-type pyramid having an equivalent cone apical angle of 139.4°. The indenters were attached to the top platen of the press and graphite powder was employed to minimize adhesion and friction between the indenters and the clay. The specimens measured 63 mm in diameter and 19 mm in thickness, and were encased in oedometer steel rings for confinement purposes. The specimens which were placed on the bottom platen of the press were raised at a constant rate of 0.5 mm/min. The load and the indenter depth were continuously recorded during each test.

4   Analysis and Discussion of Indentation Test Results

4.1   Determination of Young's Modulus

Typical loading-unloading curves obtained with the indenters are shown in Fig. 6. Although the shapes of the experimental compliance curves are similar at first sight to the general trend illustrated in Fig. 1, the particular form of the initial portions of the unloading branches, which are characterized by either increasing or constant penetration with decreasing load, prevented their use for the determination of the elastic modulus based on Eq. 9 from Sneddon's work. Instead, it was assumed, as suggested by Fischer-Cripps (2002), that the first few experimental points of the loading branches represented a truly elastic response. Thus, the initial portions of the loading branches which are referred to as "Experimental" in Fig. 7 were fitted with expressions of the form $P = k\ h^2$, where $k = 2E \tan \alpha/\pi\ (1 - v^2)$ from Eq. 9. Computed values of k and E are reported in Table 1, together with a second set of parameters derived using the same expression, but this time with the fitting applied to the whole loading branches. These are called "Analytical" in Fig. 7. In addition, because $P = k\ h^2$, the mean contact pressure $p_m = P/\pi\ a^2$ or $p_m = k\ h^2/\pi\ a^2 = k \cot \alpha^2/\pi$, where a = radius of circle of contact and $\cot \alpha = h/a$. Values of p_m are also reported in Table 1.

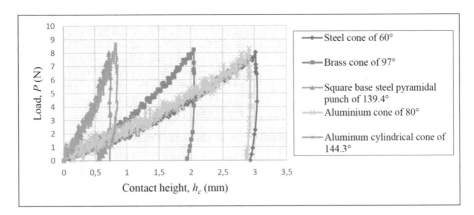

Fig. 6. Experimental compliance curves (after Ewane 2018)

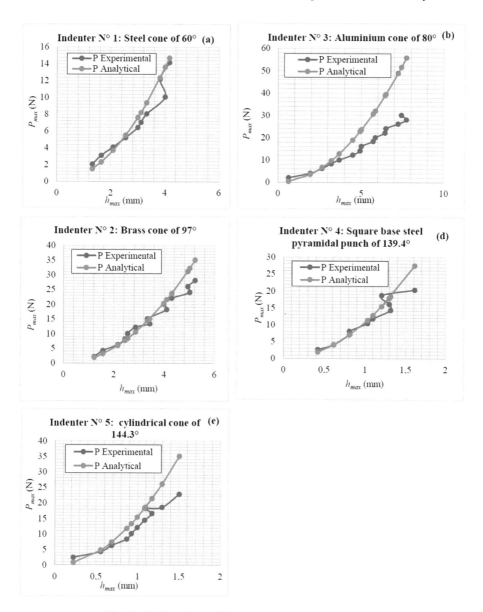

Fig. 7. Fitting of compliance curves (after Ewane 2018)

Examination of the data shown in this table indicates that (a) the computed values of Young's modulus E are much smaller than those deduced from the unconfined compression tests, and (b) the values derived for the sharper cones (i.e., 2α of $60°$, $80°$, and $97°$) are much smaller than those obtained for the blunter ones (i.e., 2α of $139.4°$ and $144.3°$). This shows that the sharper indenters induce more damage to the soil structure than the blunter indenters. As a consequence, realistic values of Young's modulus could not be determined from indentation tests with conical indenters, at least for apical angles $2\alpha \le 144.3°$.

Table 1. Values of k and E from Sneddon's approach

Indenter	Apical angle $2\alpha°$	Experimental curves			Analytical curves		
		k (kPa)	E (kPa)	p_m (kPa)	k (kPa)	E (kPa)	p_m (kPa)
Cone	60	615	1254	587	657	1340	627
Cone	80	850	1193	384	940	1320	425
Cone	97	1250	1303	311	1300	1355	384
Pyramid	139.4	10000	4357	436	10657	4650	465
Cone	144.3	12000	4553	396	15560	5903	514

4.2 Determination of Plastic Strength Parameters

For the application of the approach of Hainsworth et al. (1996), it was necessary to determine the hardness H, that is, $H = C \sigma_y$. The constraint factor C was computed based upon the results which link C to the ratio E cot $\alpha/(1 - v^2)^2 \sigma_y$ obtained by Johnson (1970). For E = 17500 kPa, $v = 0.5$, and $\sigma_y = 2 S_u = 113.4$ kPa, the ratio E cot $\alpha/(1 - v^2) \sigma_y$ ranges between 66.3 for $2\alpha = 144.3°$ and 356.4 for $2\alpha = 60°$. As a result, the constraint factor C = 3 from Johnson (1970)'s data.

The approach proposed by Hainsworth et al. (1996) was considered first for the analysis of indentation test results because it takes into account elastic and plastic response, even though the relationship $P = A* h^2$ of Eq. 15 has the same form as that derived by Sneddon (1965) for a truly elastic material. Table 2 presents the values of the parameter A* based upon Eq. 16 for E =17500 kPa, $H = 3\sigma_y = 6S_u = 340.2$ kPa, and $v = 0.5$ for undrained conditions from the unconfined compression tests. In this table are also reported the values of the mean contact pressure p_m obtained by assuming that $p_m = A* \cot^2 \alpha/\pi$, since $p_m = P/\pi a^2$. The computed values of A* are compared in Table 2 with the corresponding values of the parameter k determined from the "Analytical" curves of Fig. 7, based upon Sneddon's approach. Examination of the data indicates that although there exists a slightly better agreement for the blunter than for sharper indenters, the values of A* are on average smaller than k. Thus, the approach proposed by Hainsworth et al. (1996) fails to provide an adequate representation of the experimental observations. Comparison between the values of the mean contact pressure p_m reported in Table 2 based upon the "Analytical" curves of Fig. 7 and the hardness $H = 3\sigma_y = 340.2$ kPa also shows that p_m is on average 42% higher than H.

Values of p_m are compared in Table 3 with those obtained from the analytical curves, as well as from both the approach of Hainsworth et al. (1996) and the expanding cavity model of Johnson (1970). The data show that the computed average value of p_m obtained based on the ECM approach is 16% greater than the value of the hardness H = 340.2 kPa and 22% smaller than the corresponding value (i.e., $p_m = 483$ kPa) based on the "Analytical" curves of Fig. 7. In addition, the computed values of p_m for the blunter cones of 139.4° and 144.3° show good agreement with the value of the hardness H.

Table 2. Comparison between parameters A*, k, and mean contact pressure p_m

Indenter	Apical angle $2\alpha°$	Hainsworth et al. (1996)		Sneddon's approach	
		A* (kPa)	p_m (kPa)	k (kPa)	p_m (kPa)
Cone	60	366	350	657	627
Cone	80	732	331	940	425
Cone	97	1316	328	1300	384
Pyramid	139.4	7153	312	10657	465
Cone	144.3	9318	308	15560	514
Average:			326		483

Table 3. Comparison between p_m (kPa) values

Indenter	Apical angle $2\alpha°$	Hainsworth et al. (1996)	Sneddon's approach	ECM
Cone	60	350	627	460
Cone	80	331	425	432
Cone	97	328	384	409
Pyramid	139.4	312	465	343
Cone	144.3	308	514	333
Average:		326	483	395

Finally, if the values of the mean contact pressure p_m derived from the "Analytical" curves are considered to correspond to the response of a rigid plastic material, then slip line theory can be applied for the determination of the undrained shear strength S_u. The results which are reported in Table 4 are based on the relationship obtained by Houlsby and Wroth (1982) between the cone factor N_c and the apical angle 2α shown in Fig. 3. In addition, the values of S_u are compared with values deduced from the approaches proposed by Hainsworth et al. (1996) and the expanding cavity model of Johnson (1970). Examination of the different entries in this table shows that the average value of S_u derived from the approach proposed by Hainsworth et al. (1996) agrees well with the value of $S_u = 56.7$ kPa obtained from the unconfined compression tests. However, such result is caused by the fact that, as the second term in the right-hand side of Eq. 16 is very small compared to the first term, then the parameter A* becomes approximately equal to $H \pi \tan^2\alpha$. As a consequence, $p_m = H$ because $p_m = A* \cot^2\alpha$. Thus, $S_u = H/6$ from the approach of Johnson (1970). The data reported in Table 4 also show that whereas the average value of the undrained shear strength obtained from the expanding cavity model is only 16% greater than that found from the unconfined compression tests, the value deduced from application of the slip line theory is unrealistic.

Table 4. Comparison of undrained shear strengths S_u (kPa)

Indenter	Apical angle $2\alpha^\circ$	Hainsworth et al. (1996)	ECM	Houlsby and Wroth (1982)
Cone	60	58.3	76.7	131.9
Cone	80	55.2	72.0	84.4
Cone	97	54.7	68.2	66.2
Pyramid	139.4	52.0	57.2	83.0
Cone	144.3	51.3	55.5	74.7
Average:		54.3	65.8	88.0

5   Conclusions

The following conclusions are drawn on the basis of the contents of the present study:

1. Realistic values of Young's Realistic values of Young's modulus could not be obtained from the initial portions of the unloading branches of the indentation tests, due to either increasing or constant penetration depth with decreasing load.
2. Application of the elastic solution to the initial portions of the loading branches of the compliance curves showed that the clay suffered severe damage from the beginning of the indentation tests. Such damage which was most severe with the sharp indenters resulted in lower values of Young's modulus.
3. Application of the approaches suggested by Johnson (1970) and Hainsworth et al. (1996) did not allow obtaining a reasonable fitting to the experimental compliance curves. Again, the cause is linked to the severe damage experienced by the clay, especially with the sharper indenters. This notwithstanding, the two approaches permitted the determination of reasonable values for the undrained shear strength.
4. Application of the slip line theory resulted in unreasonable overestimation of the undrained shear strength.
5. Because indentation tests with sharp indenters caused considerable damage to the clay structure, resulting in reduced values of Young' modulus, further investigations should be carried out using either much blunter and spherical indenters, or flat-ended cylindrical punches.

References

American Society for Testing and Materials-ASTM: Standard practice for instrumented indentation testing, E2546-15. ASTM International, pp. 1–24 (2007)

American Society for Testing and Materials-ASTM: Instrumented indentation testing: a Draft ASTM Practice. ASTM International. Standardization News, October 2003

Atkins, A.G., Tabor, D.: Plastic indentation in metals with cones. J. Mech. Phys. Solids **13**(3), 149–164 (1965)

Barquins, M., Maugis, D.: Adhesive contact of axisymmetric punches on an elastic half space: the modified Hertz-Huber's stress tensor for contacting spheres. Journal de Mécanique Théorique et Appliquée **1**(2), 331–357 (1982)

Battacharya, A.K., Nix, W.D.: Finite element simulation of indentation experiments. Int. J. Solids Struct. **24**(12), 141–147 (1988)

Bishop, R.F., Hill, R., Moh, N.F.: The theory of indentation and hardness tests. Proc. Phys. Soc. **57**(3), 147–159 (1945)

Boussinesq, J.: Application des potentiels à l'étude de l'équilibre et du mouvement des solides élastiques: des notes étendues sur divers points de physique mathématique et d'analyse. Gauthier-Villars, Paris, 736 p. (1985)

Cerruti, V.: Ricerche intorno all'equilibrio dei corpi elastici isotropi. Atti Accad. Naz. Lincei Memorie Serie III **XII**, 81–123 (1882)

Cheng, Y.-T., Li, Z.: Hardness obtained from conical indentations with various cone angles. J. Mater. Res. **15**(12), 2830–2855 (2000)

Chitkara, N.R., Butt, M.A.: Numerical construction of axisymmetric slip-line fields for indentation of thick blocks by rigid conical indenters and friction at the tool-metal interface. Int. J. Mech. Sci. **34**(11), 849–862 (1992)

Dugdale, D.S.: Cone indentation experiments. J. Mech. Phys. Solids **2**(4), 265–277 (1954)

Dugdale, D.S.: Wedge indentation experiments with cold-worked metals. J. Mech. Phys. Solids **2** (1), 14–26 (1953)

Ewane, M.-S.: Essais d'indentation sur un sol de la mer Champlain. Ph.D. thesis. École Polytechnique de Montréal, Montréal, Québec, Canada (2018)

Ewane, M.S., Silvestri, V., James, M.: Indentation of a sensitive clay by a flat-ended circular punch. J. Geotech. Geoenviron. Eng. (2018). https://doi.org/10.1007/s10706-018-0561-4

Fischer-Cripps, A.C.: Introduction to Contact Mechanics, 2nd edn. Springer, New York (2007)

Fischer-Cripps, A.C.: Nanoindentation. Springer, New York (2002)

Giannakopoulos, A.E., Larsson, P.L., Vestergaard, R.: Analysis of Vickers indentation. Int. J. Solids Struct. **31**(19), 2679–2708 (1994)

Guha, S., Sangal, S., Basu, S.: Numerical investigations of flat punch molding using a higher order strain gradient plasticity theory. Int. J. Mater. Form. **7**(4), 459–467 (2014)

Hainsworth, S.V., Chandler, H.W., Page, T.F.: Analysis of nanoindentation load displacement loading curves. J. Mater. Res. **11**, 1987–1995 (1996)

Hill, R.: The Mathematical Theory of Plasticity. Oxford University Press, London (1950)

Hill, R., Lee, E.H., Tupper, S.J.: The theory of wedge indentation of ductile materials. Proc. R. Soc. Lond. A **188** (1947). https://doi.org/10.1098/rspa.1947.0009

Hirst, W., Howse, M.G.J.W.: The indentation of materials by wedges. Proc. R. Soc. Lond. A **311** (1969). https://doi.org/10.1098/rspa.1969.0126

Houlsby, G.T., Wroth, C.P.: Direct solution of plasticity problems in soils by the method of characteristics. In: Proceedings of the 4th International Conference on Numerical Methods in Geomechanics, Edmonton, vol. 3, pp. 1059–1071 (1982)

Hu, Z., Lynne, K., Delfanian, F.: Characterization of materials' elasticity and yield strength through micro/nano-indentation testing with a cylindrical flat-tip indenter. J. Mater. Res. **30** (04), 578–591 (2015)

International Organization for Standardization-ISO: Test method for metallic and non-metallic coatings, 14577-4 (2007)

International Organization for Standardization-ISO: Instrumented indentation test for hardness and materials parameters, 14577-1 (2002)

Johnson, K.L.: Contact Mechanics. Cambridge University Press, Cambridge (1985)

Johnson, K.L.: The correlation of indentation experiments. J. Mech. Phys. Solids **18**(2), 115–126 (1970)

Leroueil, S., Tavenas, F., Le Bihan, J.-P.L.: Propriétés caractéristiques des argiles de l'est du Canada. Can. Geotech. J. **20**(4), 681–705 (1983)

Lockett, F.J.: Indentation of a rigid-plastic material by a conical indenter. J. Mech. Phys. Solids **11**(5), 345–355 (1963)

Love, A.E.H.: Boussinesq's problem for a rigid cone. Q. J. Math. **10**(1), 161–175 (1939)

Marsh, D.: Plastic flow in glass. Proc. R. Soc. Lond. A **279**(1378), 420–435 (1964)

Meyer, E.: Untersuchungen über Härteprüfung und Härte Brinell Methoden. Z. Ver. deut. Ing. **52**, 66–74 (1908)

Mulhearn, T.: The deformation of metals by Vickers-type pyramidal indenters. J. Mech. Phys. Solids **7**(2), 85–88 (1959)

Prandtl, L.: Über die Härte plastischer Körper. Nachrichten von der Gesellschaft der Wissenschaften zu Göttingen, Mathematisch-Physikalische Klasse, pp. 74–85 (1920)

Riccardi, B., Montanari, R.: Indentation of metals by a flat-ended cylindrical punch. Mater. Sci. Eng. A **381**, 281–291 (2004)

Samuels, L., Mulhearn, T.: An experimental investigation of the deformed zone associated with indentation hardness impressions. J. Mech. Phys. Solids **5**(2), 125–134 (1957)

Shield, R.T.: On the plastic flow of metals under conditions of axial symmetry. Proc. R. Soc. Lond. A **233**(1193), 267–287 (1955)

Sneddon, I.N.: The relation between load and penetration in the axisymmetric Boussinesq problem for a punch of arbitrary profile. Int. J. Eng. Sci. **3**(1), 47–57 (1965)

Sneddon, I.N.: Boussinesq problem for rigid cone. Math. Proc. **44**(4), 492–507 (1948)

Tabor, D.: The Hardness of Metals. Oxford University Press, London (1951)

Wang, Z., Basu, S., Murthy, T.G., Saldana, C.: Modified expansion theory formulation for circular indentation and experimental validation. Int. J. Solids Struct. **97–98**, 129–136 (2016)

Yu, W., Blanchard, J.P.: An elastic-plastic indentation model and its solutions. J. Mater. Res. **11**(9), 2358–2367 (1996)

Zeng, K., Chiu, C.H.: An analysis of load-penetration curves from instrumented indentation. Acta Mater. **49**(17), 3539–3551 (2001)

Evaluation of Dolime Fine Performance in Mitigating the Effects of an Expansive Soil

Ahmed Hisham, Shehab Wissa[✉], and Ayman Hasan

Soil Mechanics and Foundations Department, Public Works Department,
Cairo University, Giza, Egypt
tiba.ce@gmail.com, shehabagaiby@cu.edu.eg,
aymanhll88@gmail.com

Abstract. Expansive soils are classified as problematic soils that expand when in contact with water and shrink after drying out. The soils reactivity with water is due to the presence of clay minerals that react with water such as montmorillonite. Given the geotechnical problems associated with the expansion and shrinkage behavior of expansive soils, it is necessary to treat such soils before constructing on it. Mixing the soils with additives is considered one of the main treatment methods that has been used to reduce the expansion capabilities of these soils rendering them safe to construct on and remain stable. Dolime fine; that is obtained from crushing dolomite stone; has a great potential to be used as an additive to treat expansive soils, the reason for that comes from it being composed of a percent of calcium oxide (CaO) which is known for being a binding agent that can stabilize expansive soils. In the presented experimental study, dolime chips were brought from Erbil city (northern Iraq), while bentonite was brought from Samawa city (southern Iraq), as for natural soil it was brought from the marshes of Basra city (southern Iraq) for investigation. To assess the effectiveness of dolime fine in stabilizing expansive soils, a series of laboratory tests were conducted on an artificial expansive soil; that is composed of 75% bentonite and 25% natural clay; that was mixed with dolime fine passing through sieve No. 40. The series of experimental tests conducted on the dolime fine-expansive soil mixture include unconfined compressive strength tests (UCS); compaction tests; swelling tests; and California bearing ratio tests (CBR). Through the results of these tests, a conclusion can be reached to how much of an effect does the mixing of the dolime fine with an expansive soil have on the expansion ability of the soil under study.

1 Introduction

Expansive soil is a type of clayey soil having montmorillonite minerals which expand significantly when in direct contact with water and shrink when the water dries out. Typically, the shear strength of such soils is very low which can alternate the swell-shrink behavior of the soil which subsequently may damage lightly loaded structures that are constructed on top of it. Expansive soils are treated as problematic soil for construction (Nilson and Miller 1992; Gourley et al. 1993; Rao et al. 2008; Sabat and Pati 2014). Therefore, it is necessary to improve the geotechnical properties (mainly shear strength and deformability) of swelling soils to mitigate potential damages to structures.

© Springer Nature Switzerland AG 2020
L. Hoyos and H. Shehata (Eds.): GeoMEast 2019, SUCI, pp. 19–28, 2020.
https://doi.org/10.1007/978-3-030-34206-7_2

Traditionally, complete or partial replacement of expansive soils with non-expansive engineered fill is used to lower the adverse effects of soil volume changes for structures, roads, and utilities. Alternatively, other methods may be used to reduce soil swelling potential including chemical treatments with additives such as cement or lime or fly ash (Jan et al. 2015). Stabilization is one of the techniques that is widely used to improve the geotechnical properties of expansive soils. The treated expansive soil cab be non-contaminated as summarized in numerous studies in the geotechnical literature carried out by Cocka 2001, Kalkan and Akbulut 2004, Sabat and Das 2009, Ogbonnaya and Illoabachie 2011, Sabat and Nanda 2011, Moses and Saminu 2012, Sabat 2012, Mir and Sridharan 2013, Sabat 2013, Sabat and Pradhan 2014, Ashango and Patra 2014, Sabat and Nayak 2015, Kulkarni et al. 2016, and Sabat and Mohanta 2016. Or stabilization can be utilized with contaminated soils as studied by Sabat and Mohanta (2017); who proposed using dolime fines to improve the strength properties and swelling behavior of expansive soil that is artificially contaminated with diesel. So dolime fines can be utilized to stabilize contaminated and non-contaminated expansive soil.

Dolomite chips are required for different industrial processes, the chips are obtained by crushing dolomite stones where during crushing a solid waste is produced called *dolime fines* (Sabat and Mohanta 2015). Dolime fines have a high percentage of calcium oxide (CaO). The utilization of dolime fines has been recommended by (IRC: 88-1984) as a binding agent that can replace pure lime (Shahu et al. 2013). Sabat and Mohanta (2015) used dolime fines and other mixtures to study the strength and durability characteristics of stabilized red mud cushioned expansive soil whereas Shreyas (2017) used dolime fines to stabilize black cotton soil that is known as expansive soil. The effects of mixing dolime fines with an expansive soil were studied by Golakiya and Savani (2015) and Sabat and Mohanta (2017), where it was observed that mixing soils with dolime fines yields to a decrease in maximum dry density (MDD) and swelling pressure, and on the contrary to an increase in optimum moisture content (OMC), unconfined compressive strength (UCS) and California bearing ratio (CBR).

2 Methodology

Bentonite material; brought from Samawa city (southern Iraq); was mixed with natural clay soil from the marshes of Basra city (southern Iraq) in order to make an expansive soil mix, the mass of the resulting expansive soil comprises 75% bentonite and 25% natural clay, by weight of the soil. Dolomite stones; brought from Erbil city (northern Iraq); were crushed to dolime fines and then passed from sieve No. 40 (0.425 mm). Dolime fines were added to the soil mixture at 0, 4, 8 and 12% by weight of the soil and mixed properly.

Standard Proctor compaction tests were conducted on the expansive soil mix to obtain optimum moisture content and maximum dry density. Samples were prepared for measuring the unconfined compressive strength, California bearing ratio and one-dimensional consolidation behavior of the soil mixture at OMC and MDD. The tests were conducted following ASTM procedures.

3 Analysis of Test Results

3.1 Standard Compaction Test

Standard compaction test was conducted on both natural soil and soil-dolime fine mixtures with different percentages (4%, 8%, and 12% dolime fine as a percentage of the soil weight). The test results are presented in Fig. 1 showing the relationship between the dry density of soil/soil mixtures and water content. The results indicated that the maximum dry density decreases with the increase in the dolime fine content.

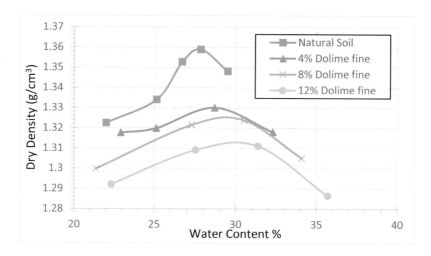

Fig. 1. Dry density versus water content for standard Proctor test

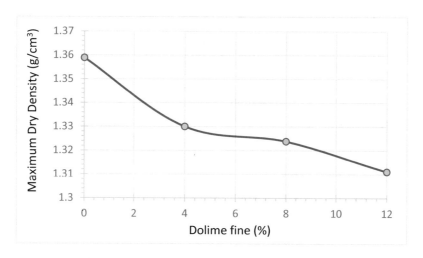

Fig. 2. Maximum dry density versus dolime fine (%)

Figure 2 shows the relationship between the maximum dry density and the dolime fines, where the maximum dry density value decreases from 1.36 g/cm^3 to 1.31 g/cm^3

with the increase of dolime fines from 4 to 12%. Furthermore, the increase in the dolime fine content leads to a corresponding increase in the values of the moisture content. The optimum moisture content increases from 27.8% to 31.4% with the increase in the added dolime fines from 0% to 12% as shown in Fig. 3.

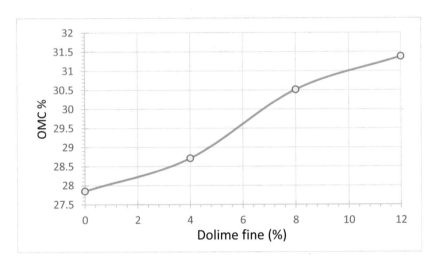

Fig. 3. Optimum moisture content versus dolime fine content

3.2 California Bearing Ratio Test (CBR)

The CBR tests were conducted on samples from natural soil condition and from the soil mixed with 4%, 8%, and 12% dolime fine by the weight. For the soil specimens mixed with dolime fine, the soil mixture resistance to penetration increases with the increase in the dolime fine content as shown in Fig. 4. CBR value for the current study gradually

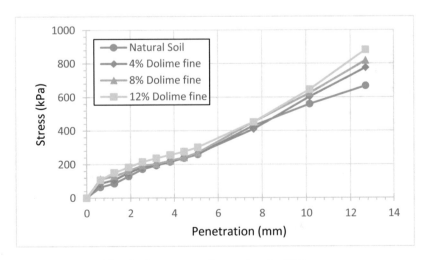

Fig. 4. Stress versus Penetration for CBR test

increases when the dolime fine is added to the swelling soil, it increases by 6.25%, 12.5%, and 25% for 4%, 8%, and 12% dolime fine, respectively; as illustrated in Fig. 5.

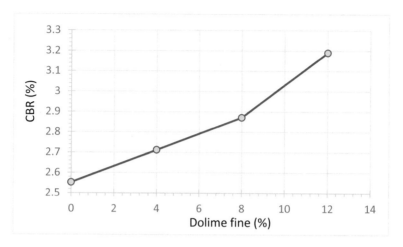

Fig. 5. CBR values versus dolime fine content for CBR test

3.3 Unconfined Compressive Strength (UCS)

Specimens of swelling clay were mixed with 0%, 4%, 8%, and 12% dolime fine by weight. All samples were mixed at its OMC and MDD. Figure 6 presents the stress-strain-strength response for all the tested soil mixtures. Figure 7 shows the increase of the UCS values with the percentage of the mixed dolime fines for the soil under study. Unconfined compressive strength increased by 60%, 90%, and 105% for dolime fine contents 4%, 8%, and 12% of soil weight, respectively.

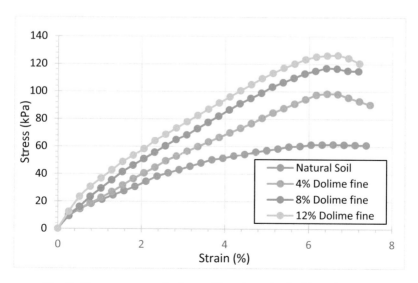

Fig. 6. Stress versus strain for swelling clay mixed with dolime fine.

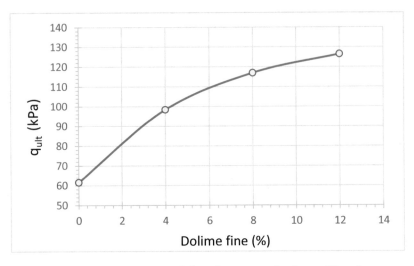

Fig. 7. UCS values versus dolime fine content for the swelling clay

3.4 One-Dimensional Swelling Test

For evaluating the swelling behavior of the expansive soil, one-dimensional swell tests were performed according to ASTM D4546-14 on both treated and untreated samples. From the results of one-dimensional consolidation (swelling pressure tests on confined samples using an oedometer cell), the swelling curves have been plotted as percent swell strain versus swelling pressure, as shown in Fig. 8. Figure 9 illustrates the swelling pressure results for soil mixed with 0%, 4%, 8%, and 12% dolime fine by weight. The increase in dolime fine content results in a decrease in both swelling pressure and corresponding strain. Figure 10 presents the variation of average swelling strain with dolime fine contents, it can be concluded that the swelling strain decreases with the increase in the dolime fines.

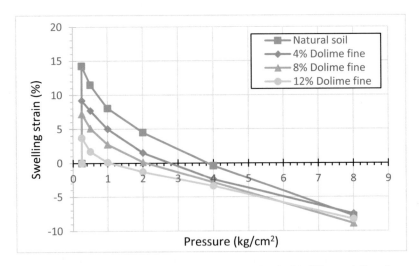

Fig. 8. Swelling strain versus pressure for swelling clay mixed with different dolime fine content

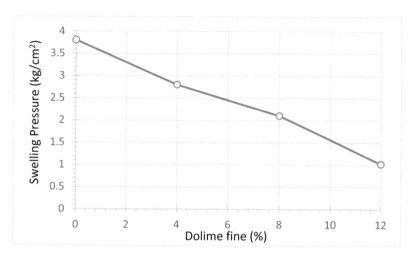

Fig. 9. Swelling pressure values versus dolime fine content for the swelling clay

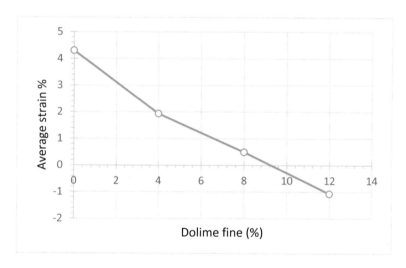

Fig. 10. Average strain versus dolime fine content

4 Conclusion

An experimental study is conducted to investigate the improvement in swelling characteristics of expansive soil due to mixing with dolime fines. A comprehensive laboratory testing program is under taken where dolime fines are added to natural clay from Samawa city (southern Iraq) with different percentages: 4, 8, and 12 by weight, to study the effect of dolime fines on the measurements of maximum dry density; optimum moisture content; unconfined compressive strength; California bearing ratio; and the swelling pressure of the expansive soil. The results were as following:

1. The maximum dry density of swelling soil–dolime fine mixture is reduced by 2%, 2.5%, and 3.5% for dolime fine contains 4%, 8%, and 12% by the weight of soil, respectively. This can be attributed to the relative light weight of the dolime fine weight compared to the weight of soil, and the large area occupied by it. Therefore, increasing dolime fine content leads to a reduction in the density of the mixture.
2. Optimum moisture content is increased by 3%, 9.5%, and 12.7% for dolime fine contains 4%, 8%, and 12% by the weight of soil, respectively. This may be due to the lack of interaction between the granules of dolime fines and water compared to attributed soil.
3. CBR values have gradually increased when dolime fines are added to swelling soil. CBR value is improved by 6.25%, 12.5%, and 25% for dolime fine contents of 4%, 8%, and 12% by the weight of soil, respectively. These increments can be attributed to the interaction between water (used to submerge samples for 4 days) and calcium oxide [CaO] presents in dolime fine. This interaction results calcium hydroxide [Ca(OH)$_2$] that is considered as solid cohesive material.
4. The values of UCS of the swelling soil – dolime fine mixture are improved by 60%, 90%, and 105% for dolime fine contents 4%, 8%, and 12% by the weight of soil, respectively. The UCS value for the current study has gradually increased when the dolime fine is added to swelling soil.
5. Addition of dolime fine to expansive clay specimens caused a considerable reduction in the swell strain and swelling pressure.
   As for the consolidation test results, mixing dolime fine with expansive soil causes a large reduction in the swelling pressure by 26%, 45%, and 73% when the swelling soil contains dolime fine 4%, 8%, and 12% by the weight of soil, respectively.
6. The swelling strain for the swelling soil – dolime fine mixture is reduced by 55%, 88%, and 125% for dolime fine contents of 4%, 8%, and 12% by the weight of soil, respectively. The decrease of swelling strain and swelling pressure occurs due to similar reason as discussed in CBR test results above.
7. The optimal ratio for the current study of the mixture of dolime fine that required for achieving a decrease in swelling properties occurs when swelling soil contains 12% of dolime fine by the weight of soil.
8. CBR, UCS, Swelling Strain and Pressure, MDD, and OMC values for swelling soil mixed with dolime fine can be summarized in Table 1.

Table 1. Summary of laboratory test results

Dolime Fine Content %	Standard Proctor Test		Improving CBR %	Improving UCS %	Decreasing Swelling Pressure %	Decreasing Swelling Strain %
	Decreasing MDD %	Increasing OMC %				
0	0	0	0	0	0	0
4	2	4	6.25	60	26	55
8	2.5	8	12.5	90	45	88
12	3.5	12	25	105	73	125

Where MDD: Maximum Dry Density; OMC: Optimum Moisture Content; CBR: California Bearing Ratio; and UCS: Unconfined Compressive Strength

References

Al-Omari, R.R., Hamodi, F.J.: Swelling resistant geogrid—a new approach for the treatment of expansive soils. Geotext. Geomembr. **10**(4), 295–317 (1991)

Ashango, A., Patra, N.: Static and cyclic properties of clay subgrade stabilized with rice husk ash static and cyclic properties of clay subgrade stabilized with rice husk ash. Int. J. Pavement Eng. **10**(15), 906–916 (2014)

Cokca, E.: Use of class c fly ashes for the stabilization of an expansive soil. J. Geotech. Geoenviron. Eng. **127**(7), 568–573 (2001)

Golakiya, H.D., Savani, C.D.: Studies on geotechnical properties of black cotton soil stabilized with furnace dust and dolomitic lime. Int. Res. J. Eng. Technol. **2**, 810–823 (2015)

Gourley, C.S., Newill, D., Schreiner, H.D.: Expansive soils: TRL's research strategy. In: Proceedings of 1st International Symposium on Engineering Characteristics of Arid Soils (1993)

Indian Roads Congress (IRC). Recommended practice for lime fly ash stabilized soil base/subbase in pavement construction. Indian Roads Congress, New Delhi (1984)

Jan, U., Sonthwal, V.K., Duggal, A.K., Rattan, E.J.S., Irfan, M.: Soil stabilization using shredded rubber tire. Int. Res. J. Eng. Technol. **2**(9), 741–744 (2015)

Kalkan, E., Akbulut, S.: The positive effects of silica fume on the permeability, swelling pressure and compressive strength of natural clay liners. Eng. Geol. **73**(1–2), 145–156 (2004)

Kulkarni, A., Sawant, M., Battul, V., Shindepatil, M., Aavani, P.: Black cotton soil stabilization using bagasse ash and lime. Int. J. Civ. Eng. Technol. **7**(6), 460–471 (2016)

Mir, B.A., Sridharan, A.: Physical and compaction behavior of clay soil–fly ash mixtures. Geotech. Geol. Eng. **31**(4), 1059–1072 (2013)

Moses, G.K., Saminu, A.: Cement kiln dust stabilization of compacted black cotton soil. Electron. J. Geotech. Eng. pp. 825–836, 17 January 2012

Nilson, J., Miller, D.: Expansive Soils: Problems and Practice in Foundation and Pavement Engineering. Wiley, New York (1992)

Ogbonnaya, I., Illoabachie, D.: The potential effect of granite dust on the geotechnical properties of abakaliki clays. Cont. J. Earth Sci. **1**(6), 23–30 (2011)

Rao, M.R., Rao, A.S., Babu, R.D.: Efficacy of cement-stabilized fly ash cushion in arresting heave of expansive soils. Geotech. Geol. Eng. **26**(2), 189–197 (2008)

Sabat, A.K.: Statistical models for prediction of swelling pressure of a stabilized expansive soil. Electron. J. Geotech. Eng. **G**(17), 837–846 (2012)

Sabat, A.K.: Prediction of California bearing ratio of soil stabilized with lime and quarry dust using an artificial neural network. Electron. J. Geotech. Eng. **18**, 3261–3272 (2013)

Sabat, A.K., Mohanta, S.: Efficacy of dolime fine stabilized red mud-fly ash mixes as subgrade material. ARPN J. Eng. Appl. Sci. **10**, 5918–5923 (2015)

Sabat, A.K., Mohanta, S.: Performance of limestone dust stabilized expansive soil-fly ash mixes as construction material. Int. J. Civ. Eng. Technol. (IJCIET) **7**(6), 482–488 (2016)

Sabat, A.K., Mohanta, S.: Unconfined compressive strength of dolime fine stabilized diesel contaminated expansive soil. Int. J. Civ. Eng. Technol. **8**(1), 1–8 (2017)

Sabat, A.K., Pati, S.: A review of literature on stabilization of expansive soil using solid wastes. Electron. J. Geotech. Eng. **19**(Bund U), 6251–6267 (2014)

Sabat, A.K., Das, S.: Design of low volume rural roads using lime stabilized expansive soil – quarry dust mixes subgrade. Indian Highways **9**(23), 21–27 (2009)

Sabat, A.K., Nanda, R.: Effect of marble dust on strength and durability of rice husk ash stabilized expansive soil. Int. J. Civ. Struct. Eng. **4**(1), 939–948 (2011)

Sabat, A.K., Nayak, R.: Evaluation of fly ash- calcium carbide residue stabilized expansive soil as a liner material in an engineered landfill. Electron. J. Geotech. Eng. **20**(15), 6703–6712 (2015)

Sabat, A.K., Pradhan, A.: Fiber reinforced–fly ash stabilized expansive soil mixes as subgrade material in flexible pavement. Electron. J. Geotech. Eng. **19**, 5757–5770 (2014)

Shahu, J.T., Patel, S., Senapati, A.: Engineering properties of copper slag–fly ash–dolime mix and its utilization in the base course of flexible pavements. J. Mater. Civ. Eng. **25**(12), 1871–1879 (2012)

Shreyas. K.: Stabilization of black cotton soil by admixtures. Int. J. Adv. Res. Sci. Eng. **6**(8) (2017). www.igarse.com

Effect of Low-Plastic Fines Content on the Properties of Clean Sand

Abdallah I. Elgendy, Shehab S. Agaiby[(✉)], and Manal A. Salem

Faculty of Engineering, Cairo University, Giza, Egypt
abdallahibrahimelgendy@gmail.com,
shehabagaiby@cu.edu.eg, manalasalem@eng.cu.edu.eg

Abstract. Physical properties of clean sand and pure clay are well established with different theories in the geotechnical literature; however, the properties of sand with fines soils are still not well determined and not fully understood. To address such matter, a series of laboratory tests were performed to investigate the effects of low-plastic silty/clayey fines on the engineering and mechanical properties of clean sand. Poorly graded clean sand was mixed with fines content percentages of 0%, 10%, 20%, and 40%. The results indicate that the shear strength parameters of tested sand measured using consolidated drained triaxial tests decrease with the increase of fines content. The internal friction angle dropped from 35° at 0% fines content to 26.86° at 40% fines content. As for the sand deformability, the modulus of elasticity (E_{50}) measured at a confining pressure of 100 kPa dropped from a value of 12823 kPa at 0% fines content to 5720 kPa at 40% fines content. The void ratio decreased as fines content increased up to 15%, then increased dramatically with further increase in fines content. Finally, to investigate the potential of using sand with fines as backfill for mechanically stabilized earth walls, the effect of fines content on interface properties between sandy soils and geogrid is also investigated by collecting data from literature and conducting regression analyses to determine the interface strength reduction factor (R_{inter}) to be used in the numerical modeling of such walls.

1 Introduction

Natural siliceous sand consists of fines and sand particles with different portions. The percentage of fines directly affects sand properties. Shear strength (effective friction angle and cohesion) and compaction characteristics of a specific soil are considered as major parameters for conducting stability analysis for almost every earthwork. Typically in construction projects, local backfill may be preferable rather than clean sand due to the high cost of replacement with clean sand and possible environmental effects. Many studies have adequately assessed the soil behavior of clean sand and pure clay; however, limited studies have thoroughly studied the behavior of sandy soil with different fines content as it is not easy to predict the response of the hybrid soil depending on the established data of clean sand (Phan et al. 2016).

© Springer Nature Switzerland AG 2020
L. Hoyos and H. Shehata (Eds.): GeoMEast 2019, SUCI, pp. 29–40, 2020.
https://doi.org/10.1007/978-3-030-34206-7_3

Najjar et al. (2015) investigated the effect of natural clay on the shear strength of clean sand. A decrease of 44% in internal friction angle of clean sand was observed in the case of mixing clean sands with 40% clay. However, Phan et al. (2016) reported that the internal friction angle of clean sand would decrease by 20% in the case of mixing with 30% clay at constant relative density.

The soil shear strength basically depends on the resistance due to interlocking, friction, and cohesion among the soil particles. Accordingly, Islam et al. (2017) defined a fineness modulus (FM) as the percentage of the cumulative retained soil on sieves up to No. 100 and limited the decrease in friction angle with the increase of fines content only for FM up to 1.5; but for FM greater than 1.5, the internal friction angle is almost constant.

Not only shear strength properties would be affected but also compaction characteristics would be changed by increasing the fines content. Despite the decrease in shear strength due to the increase in the fines content, minimum void ratio requirements would be enhanced with increasing fines content up to about 18% (Phan et al. 2016).

In the current study, a series of laboratory tests were employed to investigate the effects of fines content on shear strength properties, soil stiffness, and void ratio of sandy soil. This study is a part of an ongoing research investigating the potential use of sands with different fines contents as backfill for mechanically stabilized earth (MSE) walls. Therefore, the effect of fines content on the interface properties between the backfill soil and geogrid used in MSE walls is also investigated by collecting data from literature and conducting regression analyses to determine the interface strength reduction factor (R_{inter}) to be used in the numerical modeling of such walls.

2   Test Material

The sand used in the current study is classified as poorly graded sand (SP) according to the Unified Soil Classification System (USCS) (ASTM D2487). Fines mixed with the sand samples has a liquid limit of 24.5% and plasticity index of 5.6%. The fines are classified as low plastic silt/clay (CL/ML) according to USCS.

Tests were performed on four sand-fines mixtures defined by dry weight: 100% sand (sample 1), 90% sand plus 10% fines (sample 2), 80% sand plus 20% fines (sample 3), and 60% sand plus 40% fines (sample 4). The measured specific gravity of clean sand is 2.69 and of fines is 2.75; hence, the specific gravity of the three mixtures ranges from 2.69 to 2.75 depending on the amount of fines in the mixture. Specific gravity values of 2.7, 2.7, and 2.71 were determined for sample 2, sample 3, and sample 4, respectively. Grain size distribution curves for different sand-fines mixtures are presented in Fig. 1.

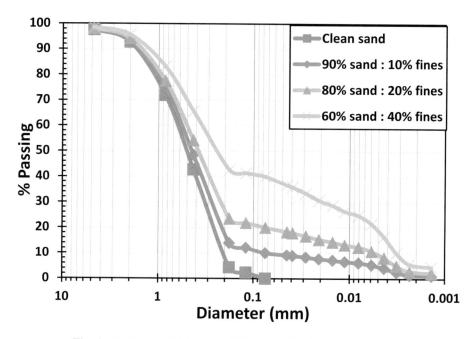

Fig. 1. Grain size distribution of different soil mixtures under study

3 Experimental Program

It is worth mentioning that there are no applicable ASTM standard procedures for determining the maximum and minimum void ratios for all investigated ranges of fines content (Phan et al. 2016). The ASTM D4254 and D4253 methods are limited to the determination of maximum and minimum void ratios with a maximum fines content of 15% in the mixture. Despite these limitations, the maximum and minimum soil density are still tested according to this specification. The ASTM D4253 and D4254 were used to test the maximum density to obtain the minimum void ratio index and the minimum density to obtain the maximum void ratio index, respectively.

A series of consolidated drained (CD) triaxial tests were performed according to ASTM D7181 to evaluate the shear strength and stiffness parameters of different sand/fines mixtures. Samples were tested at a constant relative density of 75% under confining stresses of 50 kPa, 100 kPa, and 200 kPa. The dry screening method as defined in ASTM D7181 was used during samples preparation then backpressure; with a value of 300 kPa; was added gradually to saturate the specimen, which was considered achieved when the pore-pressure coefficient "B" (Skempton 1954) was 0.95 or greater. Vertical stress, confining pressures, volume change, and vertical displacement were measured during shearing.

Primary compression elastic modulus for different mixtures calculated from one-dimensional consolidation test was evaluated according to ASTM D2435. The samples were loaded with stresses (5, 50, and 100 kPa) to record the primary compression for each stage. The initial void ratio of samples was calculated to achieve the target relative density of 75%.

4   Test Results and Discussions

4.1   Maximum and Minimum Soil Densities Test

The values of maximum density recorded were 1.74, 1.79, 1.75, and 1.69 g/cm^3 for sample (1), sample (2), sample (3), and sample (4), respectively. And minimum density values were 1.53, 1.57, 1.52, and 1.30 g/cm^3 for the same samples. The test results indicate that the increase in fines content lead to a decrease in the measured void ratio values up to an optimum value, which occurs at a critical fines content of about 15%; then a drastic increase in void ratio was observed as shown in Fig. 2.

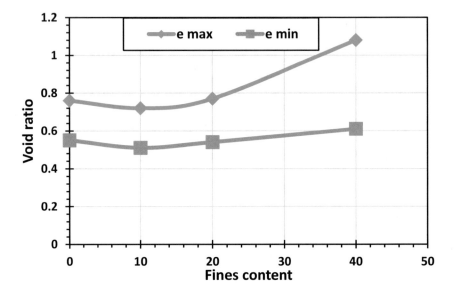

Fig. 2. Effect of fines content on void ratio index

Generally, the critical fines content corresponding to minimum void ratio changes from one soil mixture to another. It greatly depends on the summation of particles surface area which controls the natural void ratio of any soil where each particle is surrounded by air voids. In the presented study, the sand particles and corresponding fines have an initial independent void ratio before mixing. During mixing, air voids in sand samples are gradually replaced by fines until reaching a state where all sand particles are enclosed by fines (i.e., critical fines content). After which, any increase in the fines content has no effect on the voids within the sand samples but increases the void ratio of the mixture.

4.2   Triaxial Test

Shear stress and volume change versus vertical displacement were measured as presented in Fig. 3. It was observed that all soil samples exhibit the same stress-strain-strength behavior of peak and residual strength, the drop from peak to residual can be

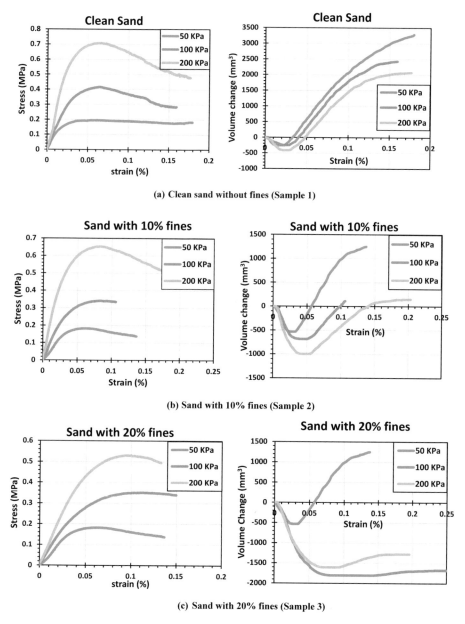

Fig. 3. Shear stress and volume change versus vertical displacement for four different mixtures under study

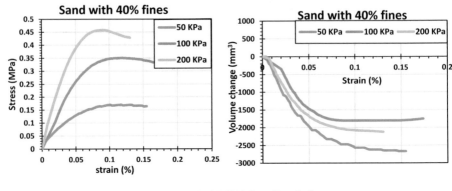

(d) Sand with 40% fines (Sample 4)

Fig. 3. (*continued*)

Table 1. Soil shear strength and stiffness parameters of different investigated soil samples

	Sample 1	Sample 2	Sample 3	Sample 4
Friction angle (ϕ')	35.02°	32.46°	29.16°	26.86°
E_{50} (at confining pressure of 100 kPa)	12823 kPa	8996 kPa	7111 kPa	5720 kPa

attributed to the orientation of soil particles along the slip surface that can lead to a volume decrease and also attributed to the increase in water content because of the dilatancy of soil particles over the shear zone. These two effects increase with the increase of the sand content (Cetin and Söylemez 2004). The difference between the peak and residual was observed to increase with the increase of the sand content at different stress levels.

It can be observed that the soil behavior changed gradually from dense to loose sand with the increase in the fines content. Upon shearing, the clean sand sample showed strain hardening behavior and dilate while the volume change showed hardening behavior for clean sand then moved toward softening behavior with the increase of the fines content. Volume change curves indicated that clean sand compresses at small displacements, then dilates until failure was reached. This compression was noticed to increase gradually with the increase in the fines content; from approximately 5% strain (in case of clean sand) to no dilation behavior (in the case of 40% fines).

The gradual change of sand to looser sand behavior can be also observed from shear strength parameters. The internal friction angle ($\phi°$) decreased with the increase of fines content; however, no cohesion (c) value was measured at the different fines content because of the non-plastic fines behavior. The stress curves show that clean sand failed at low strains, which increased gradually with the increase of fines content. In addition, the soil strength for any given confining pressure decreased with the increase of fines content and this means that the Mohr-Column envelope failure plane moved downwards causing a decrease in shear strength parameters (internal friction angle ϕ, cohesion c) and soil stiffness (E_{50}) as shown in Table 1. The internal friction angle dropped from 35.0° (in case of clean sand) to 26.86° (in case of 40% fines). Figure 4 shows the decrease of internal friction angle with fines content increase. An empirical relationship shown below can be estimated for internal friction angle prediction with fines content as follows:

$$\phi° = -0.2046\,(\%F.C) + 34.456°\qquad(1)$$

where (ϕ) is the internal friction angle in degree and %F.C is the percentage of fines content. This relationship is only valid for clean sand with $\phi = 34°–36°$ mixed with low plastic fines.

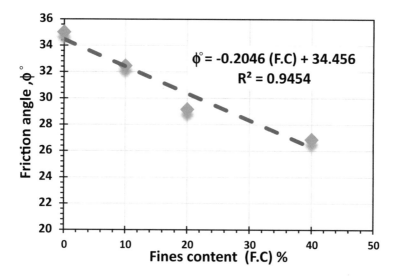

Fig. 4. Change of effective friction angle with fines content

It is worth mentioning that the B value, which represents the degree of saturation of tested soil, plays a vital role. All tests are performed for B value over 95% the wide range of B value effect can be observed in volume change for sand with 20% fines and sand with 40% fines. For sand with 20% fines, the volume change of the test at 100 kPa is lower than that at 200 kPa which can be attributed to a higher degree of saturation (high B value) for 100 kPa sample. The same explanation could be for sand with 40% fines at 50 kPa confining pressure sample.

4.3   One Dimensional Consolidation Test

The coefficient of volume compressibility (m_v) was determined for the stress range 50 to 100 kPa. The values of E_{oed} obtained were approximately the same of E_{50} from the triaxial test (Abdelmawla et al. 2014) as shown in Table 2.

Table 2. Modulus of elasticity of different samples under study due to primary compression.

	Sample 1	Sample 2	Sample 3	Sample 4
E_{oed} (kPa)	13248	8898	6920	5522

5   Interface Strength Reduction Factor (R_{inter}) with Geo-Grid

This study is part of an ongoing research investigating the potential use of sands with different fines contents as backfill for mechanically stabilized earth (MSE) walls. Therefore, the effect of fines content on the interface properties between the backfill soil and geogrid used in MSE walls is also investigated. Evaluating soil-geogrid reinforcement interaction is essential for accurate analysis and design of reinforced soil systems. Soil shear strength properties are reduced by a reduction factor on the geogrid interface (R_{inter}), which is defined as the ratio between soil-geogrid shear strength and soil-soil shear strength. Over the past years, numerous experimental studies with direct shear and pullout tests have been conducted to improve the understanding of geogrid-soil interaction. The geogrid embedded in granular soil can resist pullout force by friction and bearing resistances (Sukmak et al. 2016). Moreover, few numerical studies have been undertaken to validate experimental results (Abdelmawla et al. 2014; Hegde and Roy 2018; Touahmia et al. 2018), and other studies have tried to predict the pullout behavior of geogrid embedded in granular soil theoretically (Moraci et al. 2007).

For numerical pullout simulation for sandy soil, some studies assumed a value of 0.67 for R_{inter} as per FHWA-NHI-00-043, 2001 which is only valid for clean sand with fines content up to 15%, uniformity coefficient (Cu) should be greater than or equal 4, and plasticity index (PI) should not exceed 6 (Abdelmawla et al. 2014). Other studies assumed that the soil-geogrid surface is rigid and thus a value of 1.0 for R_{inter} was assumed (Touahmia et al. 2018). Hegde and Roy (2018) changed the value of R_{inter} until the results matched performed pullout test. Generally, R_{inter} ranges between 0.67 and 1 for clean sands depending on soil condition, geogrid type, and test condition.

Although many studies have mentioned factors that affect soil-geogrid interface, till now there are a lot of uncertainties regarding the selection of appropriate design parameter of R_{inter} for geogrid-reinforced soil systems. These uncertainties may be due to duplicate or not take all influencing factors into account.

In this study, R_{inter} back-calculated from pullout tests and numerical simulation results reported in the literature were investigated along with its corresponding test and soil condition to study which factors have a significant effect on R_{inter}. Studied factors were soil relative density (D_r), soil internal friction angle (ϕ), soil mean particle size (d_{50}), normal stress applied during the test (σ_n), geogrid length (L), number of transverse geogrid members, and geogrid opening size (#).

It was observed that R_{inter} is a characteristic value that only depends on soil and geogrid properties. However, test conditions like normal stress affects peak and residual strength but the fraction from this strength transmitted to the geogrid interface (R_{inter}) remains constant. Internal friction angle (ϕ) and soil mean particle size (d_{50}) were observed to have much more effect on R_{inter} values. Moreover, relative density affected R_{inter} values indirectly by changing the internal friction angle (ϕ). Also, geogrid opening size (#) was observed to be the most influential factor on R_{inter} for geogrid. Thus, such factors can be summarized in three major factors: geogrid geometry (#), mean grain size of soil (d_{50}) and soil internal friction angle (ϕ).

Table 3 shows data carefully collected from previous studies in the literature (covering several sand sites from China and Poland). R_{inter} values greater than one were excluded as they seemed unrealistic to develop a relationship estimating R_{inter} value from the geogrid opening size (#), the mean grain size of soil (d_{50}), and soil internal friction angle (ϕ).

Table 3. Pullout tests data collected from different researches

Reference	R_{inter}	tan ϕ	d_{50} (mm)	Opening (mm)
Lentz and Pyatt (1988)	0.95	0.676	0.35	47
Duszyńska and Bolt (2004)	0.93	0.721	1.19	46.69
Hsieh et al. (2011)	0.93	0.748	0.64	27.24
	0.88	0.793	6.9	27.24
Shi and Wang (2013)	0.87	0.637	0.26	55.15
Wang et al. (2018)	0.73	0.675	0.25	300.66

Figure 5 shows the results of simple regression analysis for the collected data from which the following equation can be developed

$$R_{inter} = -0.0016\left(\frac{\#^{0.2}(mm)}{d_{50}(mm).\tan\phi}\right)^2 + 0.0222\left(\frac{\#^{0.2}(mm)}{d_{50}(mm).\tan\phi}\right) + 0.874 \quad (2)$$

where R_{inter} is reduction factor on the geogrid interface, # is the geogrid opening size in mm, d_{50} is mean grain size of soil in mm and ϕ is the internal friction angle of the soil in degrees.

Figure 5 shows that R_{inter} would slightly increase by increasing the fraction of $\frac{\#^{0.2}(mm)}{d_{50}(mm).\tan\phi}$ From zero up to 7, that's because this part of the curve sandy soil-geogrid interaction approached gravel-geogrid interaction behavior, high $d_{50},$ and tan ϕ, then transmitted gradually to sand behavior. The remaining part of the curve for fraction value above 7 shows the effect of fines content on R_{inter}. This relationship verifies the fact that R_{inter} value depends on soil properties more than that of the geogrids which is well established by (Hossain et al. 2011).

The above-concluded relationship is applied on different soil mixture samples to evaluate R_{inter} values with a geogrid of opening size 16×219 mm. R_{inter} values are 0.95, 0.93, 0.89 and 0.53 for sample 1, sample 2, sample 3 and sample 4 respectively.

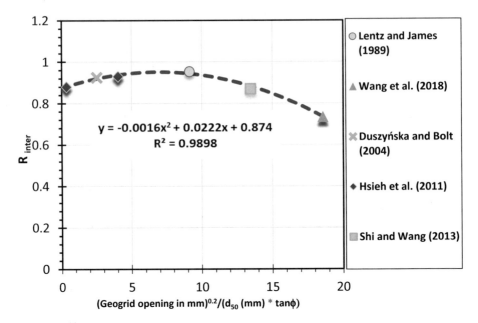

Fig. 5. Prediction of sandy soil R_{inter} value at different geogrid and mean particle size conditions

6   Conclusions

In the presented study, the effects of low plastic fines content on properties of sand were investigated by a series of laboratory tests (sieve analysis, hydrometer test, Atterberg limits, specific gravity, CD Tri-axial test, one dimensional consolidation test and maximum and minimum densities) conducted on poorly graded sand mixed with fines in different portions (0%, 10%, 20% and 40% by weight) and pullout test results data collection from the geotechnical literature. Based on the collected and measured data, the following conclusions can be derived:

- The smallest values of maximum and minimum void ratio were measured at fines contents of 15%.
- Low plastic fines have no effect on the cohesion of the sand under study.
- The soil behavior changed gradually from stiff to weak soil behavior with the increase of fines content from 0% to 40%, the internal friction angle ($\phi°$) dropped to a value equal to 75% of the initial measured value, whereas, the modulus of elasticity (E_{50}) measured at a confining pressure 100 kPa dropped to a value equal to 45% of the initial one.
- Modulus of elasticity values due to primary compression (E_{oed}) are approximately equal to the modulus of elasticity obtained from the triaxial test (E_{50}) at the same confining pressure (100 kPa).
- Interface strength reduction factor (R_{inter}) is affected by fines content which in turn affects different soil properties.

– An empirical relationship was introduced to determine the interface strength reduction factor (R_{inter}) for usage in various numerical modeling applications based on the internal friction angle of the soil, mean soil particles diameter and geogrid opening size.

Acknowledgments. Sincere thanks and appreciation are expressed to Dr. Mohamed El Kholy, Dr. Mohamed Kamal, and Eng. Islam Mamdouh for their support, valuable contribution, and guidance during conducting the experimental work.

References

Abdelmawla, A., Sherif, O., Akl, S., Alashaal, A.: Analysis of laboratory pull-out experiment on geosynthetics. In: Proceedings of AIC SGE8, Structural Engineering Department, Alexandria University, vol. 1, pp. 64–72 (2014)

ASTM D2435-2003: Standard Test Methods for One-Dimensional Consolidation Properties of Soils Using Incremental Loading. 100 Barr Harbor Drive, PO Box C700, West Conshohocken, PA 19428-2959, United States

ASTM D422-2007: Standard Test Method for Particle-Size Analysis of Soils. 100 Barr Harbor Drive, PO Box C700, West Conshohocken, PA 19428-2959, United States

ASTM D4253-2006: Standard Test Methods for Maximum Index Density and Unit Weight of Soils Using a Vibratory Table. 100 Barr Harbor Drive, PO Box C700, West Conshohocken, PA 19428-2959, United States

ASTM D4254-2006: Standard Test Methods for Minimum Index Density and Unit Weight of Soils and Calculation of Relative Density. 100 Barr Harbor Drive, PO Box C700, West Conshohocken, PA 19428-2959, United States

ASTM D7181-2011: Standard Test Method for Consolidated Drained Triaxial Compression Test for Soils. 100 Barr Harbor Drive, PO Box C700, West Conshohocken, PA 19428-2959, United States

ASTM D854-2010: Standard Test Methods for Specific Gravity of Soil Solids by Water Pycnometer. 100 Barr Harbor Drive, PO Box C700, West Conshohocken, PA 19428-2959, United States

Cetin, H., Söylemez, M.: Soil-particle and pore orientations during drained and undrained shear of a cohesive sandy silt clay soil. Can. Geotech. J. **41**(6), 1127–1138 (2004)

Duszyńska, A., Bolt, A.F.: Pullout tests of geogrids embedded in non-cohesive soil. Arch. Hydro-Eng. Environ. Mech. **51**(2), 135–147 (2004)

Hegde, A., Roy, R.: A comparative numerical study on soil-geosynthetic interactions using large scale direct shear test and pullout test. Int. J. Geosynthetics Ground Eng. **4**(1), 2 (2018)

Hossain, M.S., Kibria, G., Khan, M.S., Hossain, J., Taufiq, T.: Effects of backfill soil on excessive movement of MSE wall. J. Perform. Constructed Facil. **26**(6), 793–802 (2011)

Hsieh, C.W., Chen, G.H., Wu, J.H.: The shear behavior obtained from the direct shear and pullout tests for different poor graded soil-geosynthetic systems. J. GeoEng. **6**(1), 15–26 (2011)

Islam, T., Islam, M.A., Islam, M.S., Abedin, M.Z.: Effect of fine content on shear strength behavior of sandy soil. In: Proceedings of 14th Global Engineering and Technology Conference. BIAM Foundation, 63 Eskaton, Dhaka, Bangladesh (2017). ISBN: 978-1-925488-60-9

Lentz, R.W., Pyatt, J.N.: Pull-out resistance of geogrids in sand. Transp. Res. Rec. **1188** (1988)

Moraci, N., Cardile, G., Gioffrè, D.: A theoretical method to predict the pullout behavior of extruded geogrids embedded in granular soils. In: Proceedings of 5th International Symposium on Earth Reinforcement, Fukuoka, Japan, vol. 14, p. 16 (2007)

Najjar, S.S., Yaghi, K., Adwan, M., Jaoude, A.A.R.A.: Drained shear strength of compacted sand with clayey fines. Int. J. Geotech. Eng. 9(5), 513–520 (2015)

Phan, V.T.A., Hsiao, D.H., Nguyen, P.T.L.: Effects of fines contents on engineering properties of sand-fines mixtures. Procedia Eng. 142, 213–220 (2016)

Shi, D., Wang, F.: Pull-out test studies on the interface characteristics between geogrids and soils. EJGE 18, 5405–5417 (2013)

Skempton, A.W.: The pore-pressure coefficients A and B. Geotechnique 4(4), 143–147 (1954)

Sukmak, K., Han, J., Sukmak, P., Horpibulsuk, S.: Numerical parametric study on behavior of bearing reinforcement earth walls with different backfill material properties. Geosynthetics Int. 23(6), 435–451 (2016)

Touahmia, M., Rouili, A., Boukendakdji, M., Achour, B.: Experimental and numerical analysis of geogrid-reinforced soil systems. Arab. J. Sci. Eng. 43(10), 5295–50303 (2018)

Wang, H.L., Chen, R.P., Liu, Q.W., Kang, X., Wang, Y.W.: Soil–geogrid interaction at various influencing factors by pullout tests with applications of FBG sensors. J. Mater. Civ. Eng. 31 (1), 04018342 (2018)

Stabilization of Expansive Soil Reinforced with Polypropylene and Glass Fiber in Cement and Alkali Activated Binder

Mazhar Syed and Anasua Guharay[✉]

Department of Civil Engineering, BITS-Pilani Hyderabad, Secunderabad, India
s.mazhar785@gmail.com,
guharay@hyderabad.bits-pilani.ac.in

Abstract. Expansive black cotton soil (BCS) exhibits dual nature (swelling/shrinkage) predominantly when it is exposed to moisture fluctuation. This behavior renders the BCS unsuitable for use in geoengineering applications. The present study emphasizes the polypropylene and glass fiber based soil reinforcement with a traditional cement binder and envirosafe alkali-activated binders (AAB). Cement stabilization is one of the most popular methods for reducing swelling properties of BCS. However, the production of cement leads to the emission of greenhouse gases, which is a threat to modern society. Hence the present study aims to compare the geomechanical strength between AAB and cement binder with inclusions of various discrete fibers. AAB is generated by the reaction between an aluminosilicate precursor (Fly ash and/or GGBS) and an alkali activator solution of sodium hydroxide and sodium silicate. The water to solids ratio is maintained at 0.4 in the present study. Mineralogical and microstructural characterization are performed for both cement and AAB treated BCS as well as untreated BCS through stereomicroscope, X-ray diffraction (XRD), Fourier-transform infrared (FTIR) spectroscopy, scanning electron microscope (SEM), and energy dispersive x-ray spectroscopy (EDS). The unconfined compressive strength (UCS), indirect tensile strength (ITS), California Bearing Ratio (CBR) and consolidation characteristics of both untreated and binder treated BCS are carried out at different combinations of cement-fiber and AAB-fiber in the clay. It is observed that the proposed treatment method shows a significant improvement in geoengineering properties and aids in enhancing the shear strength and ductility properties. An addition of 5% AAB with 0.3% of polypropylene fiber reduces the plasticity and swelling pressure by 17–25%, while CBR and ITS values are increased by 28–33%. Recommendations on the practical implementation of this technique for stabilization of expansive soils are proposed based on findings of the present study.

Keywords: Stabilization of expansive soil · Alkali Activated Binder · Geotechnical characterization · Fiber · Microstructural analysis

© Springer Nature Switzerland AG 2020
L. Hoyos and H. Shehata (Eds.): GeoMEast 2019, SUCI, pp. 41–55, 2020.
https://doi.org/10.1007/978-3-030-34206-7_4

1   Introduction

Expansive clayey soil has high vulnerability to volume change and exhibits low strength bearing ratio, due to the existence of smectite group and soil moisture imbalance (Ola 1978; Ackroyd and Husain 1986; Chen 1988; Oren 2014). Low volumetric soil stability may lead to cause severe destruction of structures founded on them (Katti 1978; Petry and Little 2002; Phani kumar and Sharma 2004). To overcome the problem of shrinkage, heave, and settlement, cementitious chemical binders are used as an additive to stabilize the expansive soil (Das 2003; Sivapullaiah et al. 2009). Lime and cement are identified as the most usable traditional binders for stabilizing the expansive clay (Bell 1996; Yong et al., 1996; Chen and Wang 2006). However, production of these binders leads to the emission of carbon dioxide and nitrous oxide. Also, excessive use of binders may lead to soil shrinkage cracks due to rapid sulfate reaction (Gartner 2004; Al-Rawas 2005; Ouhadi and Yong 2008; Pourakbar et al., 2015). Industrial by-products such as fly ash, rice husk ash (RHA), Ground Granulated Blast Furnace Slag (GGBS), bagasse ash, pond ash, volcanic ash, cement kiln dust has also gained popularity in stabilizing expansive soil (Kumar et al. 2007; Tang et al. 2007; Eberemu and Sada 2013; Lin et al. 2013; Salahudeen et al. 2014; Ural 2015; Kolay et al. 2016; Miao et al. 2017; Gupta et al. 2018). This technique minimizes the disposal of industrial by-products and maintains the green sustainable environment in an economic and efficient way (Gupta et al. 2018; Mazhar et al. 2018). Numerous research is carried out to improve the geomechanical behavior of BCS with a combination of cementitious binder and fiber (Kaniraj and Gayathri 2003; Yetimoglu et al. 2005; Kuamr et al. 2007; Babu and Vasudevan 2008; Tang et al. 2007, 2010, 2014). However, limited studies are reported on the mechanism of fiber bonding with cement and fly ash based binders.

The present study proposes a method of geopolymerisation of expansive black cotton soil by alkali activated binders (AAB) prepared by blending Class-F fly ash, an industrial by-product, with a solution of sodium silicate and sodium hydroxide. The water to solids ratio is maintained at 0.4. The primary objective of this paper is to compare the microstructure and geoengineering characteristics of BCS reinforced with glass and polypropylene fiber in AAB and cement treated BCS individually. The brittle nature of BCS blended with cement and AAB mixture can be overcome through reinforcing with short discrete Polypropylene and glass fibers intrusion.

2   Materials

2.1   Black Cotton Soil (BCS)

BCS is collected in this study from Nalgonda district of Telangana State in the southern part of India. The soil is dark brown to black in color and excavated at 30 cm depth from the natural ground surface in order to avoid the collection of vegetation. The soil is classified as highly plasticity clay (CH) according to the Unified Soil Classification System. The different physical and mechanical properties of raw BCS are provided in

Table 1. Particle size distribution curves for both untreated and AAB treated BCS are shown in Fig. 1.

Table 1. Engineering properties of raw BCS

Soil properties	
Soil classification (as per USCS)	CH
Specific Gravity	2.59
Optimum Moisture Content, OMC (%)	24.5
Maximum Dry Density, MDD (g/cc))	1.65
Liquid Limit, LL (%)	62.0
Plasticity Index, PI (%)	38.0
Free Swelling Index, FSI (%)	86.0
Indirect Tensile Strength, ITS (kPa)	12.54
Unconfined Compressive Strength, UCS (kPa)	185
California Bearing Ratio, CBR (%)	
• Soaked	1.96
• Unsoaked	5.54

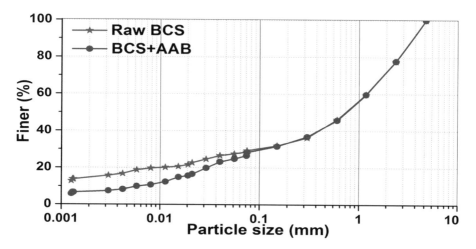

Fig. 1. Particle size distribution curve of raw BCS and AAB treated BCS

2.2   Alkali Activated Binder (AAB)

Class-F fly ash, used in the present study, is obtained from National Thermal Power Corporation (NTPC), Ramagundam city, Telangana, India. Sodium silicate solution and sodium hydroxide pellets are obtained from Hychem Chemicals Laboratories, Hyderabad. The purity of sodium hydroxide pellets is 99%. The sodium silicate solution is composed of 55.9% water, 29.4% SiO_2 and 14.7% Na_2O. The mass ratio of sodium hydroxide to sodium silicate to fly ash is 10.57:129.43:400 (Gupta *et al.* 2018).

2.3   Cement

Ordinary Portland cement of grade 53, used in this study, was sourced from JSW Cement Limited, Nandyal, Andhra Pradesh.

2.4   Fibers

Polypropylene fiber (PPF) and Glass fiber (GF) used in this study are obtained from Kanaka Durga Industries Pvt. Ltd., Hyderabad. Both fiber lengths of 12 mm are adopted in the entire test. Figure 2 shows the physical appearance of both PPF and GF.

Fig. 2. Particle Image showing discrete polypropylene and glass fiber

2.5   Preparation of Soil Sample

Raw BCS is mixed separately with 5% of AAB paste and same 5% of dry cement (with respect to the total weight of soil) satisfying optimum binder requirement for both dry cement and AAB paste. The w/s ratio for AAB paste is maintained by 0.4. Both cement and AAB mixed soil are compacted in three layers with 9 kg rammer and a fall height of 30 cm. For continuous curing, cement and AAB compacted BCS are covered with moist jute bag minimum for 24 h. Prior to random mixing of different percentages of polypropylene and glass fiber (0%, 0.1%, 0.2%, 0.3% and 0.4% by mass of BCS) individually in both cement and AAB mixed soil. The glass, polypropylene fiber reinforced with cement and AAB treated soil specimens are designated as BCS+C+GF, BCS+C+PPF, BCS+AAB+GF, BCS+AAB+PPF and where C denotes cement.

3   Results and Discussions

3.1   Microstructure Analysis

Influence of fly ash based AAB, cement and randomly oriented polypropylene and glass fibers on the microstructure analysis of expansive black cotton soil is investigated by conducting x-ray diffraction, Fourier transfer infrared spectroscopy, stereomicroscope, scanning electron microscope and energy dispersive x-ray spectroscopy.

X-Ray Diffraction (XRD)

Powder X-ray diffraction analyses are performed using a RIGAKU Ultima-IV diffractometer to identify the minerals crystallinity in BCS. Both cement and AAB treated samples are examined through CuKα rays generated at 40 mA and 40 kV. The operating 2θ range is from 0° to 80° with a step of 0.02° for 2θ values and integrated at the rate of 2 s per step. Figure 3 shows the X-ray diffraction patterns of untreated BCS and fiber reinforced cement and AAB treated BCS. The raw BCS consisted of clay minerals such as Montmorillonite (M), Quartz, (Q) and Muscovite (Ms) (Sharma et al. 2012; Miao *et al.* 2017). After addition of fly ash based AAB to BCS, negligible changes take place in the peak intensities, as is evident from the diffractograms. It is observed that the crystalline peak of Montmorillonite (M) is reduced significantly, which may be attributed towards the alteration of clay minerals (Sivapullaiah *et al.* 2009; Ural Ural 2015; Rios et al. 2015; Sekhar and Nayak 2017). The diffractogram for fiber-AAB treated BCS shows additional peaks corresponding to Quartz (Q), and Mullite (Mu) which are characteristic to the hardened AAB paste (Mia et al. 2017; Mazhar *et al.* 2018). Moreover, the XRD pattern of fiber-cement treated BCS reveals crystalline peaks of Calcite (C), and Calcium silicate hydrate (CSH) (Saride and Dutta 2016). The formation of this mineral can be related to the cementitious reaction induced between BCS and binder. A flatter portion in the Xrd patterns indicates an amorphous phase of both hardened cement and AAB paste.

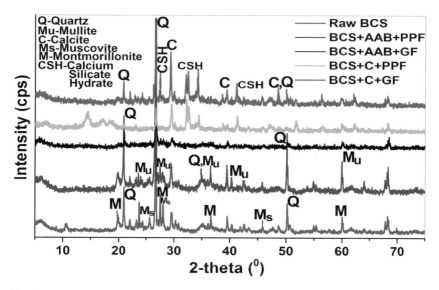

Fig. 3. Xrd pattern of untreated BCS and fiber reinforced cement and AAB treated BCS

Fourier Transforms Infrared (FTIR) Spectroscopy

Fourier transform infrared (FTIR) molecular bond spectroscopy of the untreated and treated BCS are performed using a JASCO FTIR 4200 setup with KBr pellet arrangement. Transmittance spectral range is chosen from 4000–500 cm^{-1} for all the samples.

Figure 4 shows the IR transmittance spectra of untreated BCS and fiber reinforced cement and AAB treated BCS. The spectrum curve of untreated BCS shows O-H stretching vibrations around 3616 cm^{-1} which is the general characteristics of montmorillonite (Madejova and Komadel 2001). For all selected samples, the broadband is found at 3450 cm^{-1} correspondings to O-H stretching of the hydroxyl group. Moreover, the peaks corresponding to C–H asymmetric stretching is detected between 2950–2875 cm^{-1}. The peak around 1710 cm^{-1}, representing the C=O carbonyl bond, is also detected. The bending vibration peak of =CH$_2$ group is observed at 1460 cm^{-1}. After cement and AAB-treatment, the =CH$_2$ peak in the BCS shows a chemical shift of about 10 cm^{-1}. Another main peak at 1033 cm^{-1} is attributed to Si-O-Si antisymmetric stretching, which is visible in both cement and AAB treated soil. Similarly, Al-O stretching bonds and Si-O-Al bending vibration bonds are found at 785 cm^{-1} and 527 cm^{-1} for both treated and untreated BCS, but most of them show chemical shifts, indicating impermeable nature of BCS.

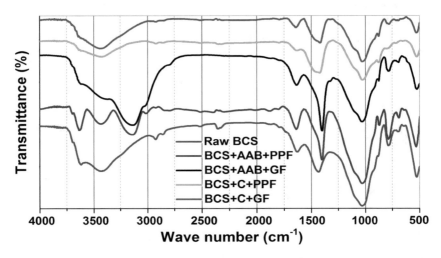

Fig. 4. FTIR spectroscopy of untreated BCS and fiber reinforced cement and AAB treated BCS

Stereomicroscopic Images

Stereomicroscopic images are visualized using an Olympus SXZ7 setup having least dimension of 20 μm. Physical surface characteristics for both untreated BCS and fiber reinforced cement and AAB treated BCS samples are captured at various magnifications. Figure 5a shows the image of untreated BCS, which consists of light red and yellow grain particles indicating the presence of iron oxide and brown colored region indicating the presence of Illite and smectite group. The presence of iron and manganese is generally associated with montmorillonite mineral of clay. Figure 5b shows a deposit of hardened AAB paste with polypropylene fiber matrix around the clay surfaces. In addition, bright and shiny regions exhibit mica from the aluminosilicate precursors and dark colored patches reflect the voids left by the evaporation of water from hardened AAB paste. Figure 5c shows the discrete glass fiber embedded in the cement treated BCS, which act as a bridge surface by holding the clay particles strongly around the fiber.

Fig. 5. Stereomicroscopy images of (a) Untreated BCS (b) Polypropylene fiber reinforced AAB treated BCS (c) Glass fiber reinforced cement treated BCS

Scanning Electron Microscope (SEM) and Energy Dispersive X-Ray Spectroscopy (EDS)

Surface morphology and elemental analysis of soil are examined using a Thermo Scientific Apreo SEM provided by FEI (Field Electron and Ion Company). Target locations are chosen randomly through Gentle beam of electromagnetic lenses which screens the surface through the small aperture by maintaining 20 kV excitation voltage with spot size 9. Energy dispersive x-ray spectra (EDS) are recorded using Aztec analyzer system provided by Oxford Instruments with a probe current of 65.4 µA at a working distance of 10 mm. Micrographs of untreated BCS and fiber reinforced AAB and cement treated BCS are presented in Fig. 6(a, b and d). The surface morphology of raw BCS reveals the flocculated flaky microstructure, which corresponds to mont-morillonite and smectite group of clay (Fig. 6a). The hollow spheres varying from small to large particles in Fig. 6b may indicate the presence of unreacted fly ash in the soil. This significant change in surface morphology of BCS may be attributed towards the formation of flocculated network with more cementitious structure (Sekhar and Nayak 2017). Figure 6c shows a discrete fiber matrix bonded with cement and AAB around the irregular aggregated clayey surfaces, which act as spatial thread groove network by interlocking the clayey particles. EDS provides the elemental composition in terms of weight percent (wt%) and atomic percent (at.%) of treated BCS. Figure 6c and d show the volumetric chemical characteristics for glass fiber reinforced AAB and cement treated BCS. Peak intensities of Silica (Si), Calcium (Ca), Oxygen (O) and Alumina (Al) become relatively stronger in AAB and cement treated BCS, which may indicate the increase in the quantity of sodium aluminosilicate and cementitious compounds. The detailed elemental analyses are tabulated in Table 2. This finding corroborates that from the SEM images.

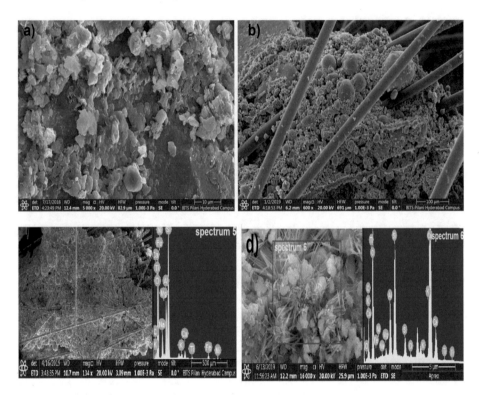

Fig. 6. SEM/EDS images of (a) Untreated BCS (b) Polypropylene fiber reinforced AAB treated BCS (c) Elemental analysis for glass fiber reinforced cement treated BCS (d) Cement treated BCS

Table 2. EDS elemental analysis for AAB and cement treated BCS

Elements/samples	BCS+AAB+GF		BCS+Cement	
	Weight %	Atomic %	Weight %	Atomic %
C	11.71	17.85	14.75	22.95
O	50.79	58.14	50.47	58.98
Na	4.77	3.80	0.18	0.15
Mg	0.56	0.42	0.45	0.35
Al	7.47	5.07	2.73	1.89
Si	19.28	12.57	6.23	4.15
K	0.76	0.36	0.16	0.08
Ca	1.77	0.81	20.93	9.76
Ti	0.48	0.18	0.18	0.07
Fe	2.07	0.68	2.43	0.81
Ni	0.34	0.10	0.21	0.07
S	0.09	0.02	1.09	0.63

3.2 Geoengineering Characteristics

A series of compaction, shear strength, indirect tensile strength and California bearing ratio tests are performed for both untreated BCS and AAB treated BCS at the different fiber content. All the soil specimens are prepared with respect to their MDD and OMC values. These experimental results are used to assess the effectiveness of binder and durability of fiber reinforced in the BCS. The details of the tests performed and the discussion of test results are given in the following sections.

Compaction
A series of standard Proctor compaction tests are performed according to ASTM D-698 standard. Figure 7 shows the MDD and OMC of untreated BCS and fiber reinforced cement and AAB treated BCS. MDD of raw BCS is 1.65 g/cc and the corresponding OMC is 24.5%, indicating highly compressible clay. Compared with the raw BCS, the addition of cement and AAB increased the MDD values from 1.65 to 1.82 g/cc; however, the OMC values decreased from 24.5 to 18.6%. The variation of dry density and moisture content can be attributed to the particle flocculation and reduction of specific surface area in the soil (Sridharan and Sivapullaiah, 2005. Sharma *et al.* 2012; Ural 2015). Thus the increase in MDD and decrease in OMC is an indicator of improving the mechanical properties of soil.

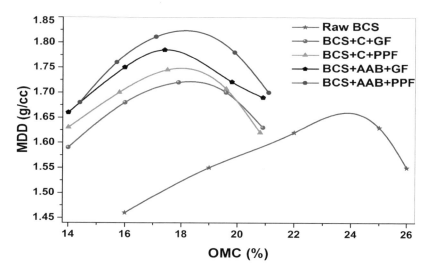

Fig. 7. Variation of MDD and OMC values of untreated BCS and fiber reinforced cement and AAB treated BCS

Consolidation and Swelling Pressure
One dimensional consolidation and constant volume swelling pressure tests are conducted in an one-dimensional consolidometer according to ASTM D-2435 and D-4546 standards. Untreated BCS, as well as cement and AAB, treated BCS samples are statically molded at MDD and OMC in a consolidation ring of 20 mm height and

60 mm diameter. Figure 8a shows the void ratio versus logarithmic effective stress (*e*-log *p*) curve of treated and untreated BCS. The *e*-log *p* of raw BCS attains the highest equilibrium void ratio and the cement treated BCS attains least void ratio on saturation. Under the confined condition, as the effective stress increases, the void ratio and swelling pressure of soil decrease upon their saturation (Sivapulliah *et al.* 2009). The reduction in the settlement may attribute to interlocking particles, density and encapsulation of clayey surfaces by deposition of hardened AAB paste and cementitious materials (Vitale *et al.* 2017; Miao *et al.* 2017). Figure 8b shows the variation of swelling pressure with time for untreated BCS as well as cement and AAB treated BCS. From the time-swell curves, it is interesting to note that that the raw BCS takes maximum time to attain the equilibrium swelling pressure when compared to cement and AAB treated the soil. The main mechanisms that govern the reduction of swelling pressure of cement and AAB modified soil are morphological changes (pozzolanic reaction, flocculation, and mineralogical alteration) and physicochemical forces (interlocking density, suction, and cation exchange) (Zhao *et al.* 2008, 2013).

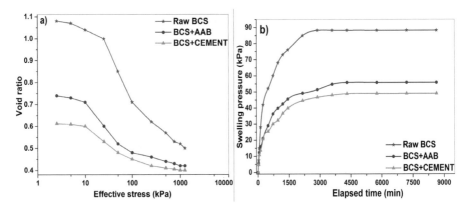

Fig. 8. Variation of (a) e-log (p) curves (b) Swelling pressure of untreated BCS and fiber reinforced cement and AAB treated BCS

Unconfined Compressive Strength (UCS)

Unconfined compressive strength tests are performed as per ASTM D-2166 standard. Treated and untreated soil samples are molded in 38 mm diameter and 76 mm height under a fixed strain rate of 1.25 mm/min. Figure 9a shows the UCS values of untreated BCS and fiber reinforced cement and AAB treated BCS. From the results, it is noticed that the addition of fiber in raw BCS does not have a significant effect on compressive strength. The inclusion of glass and polypropylene fibers with cement or AAB treated soil show significant improvement in shear resistance property. The enhancement of geomechanical characteristics of BCS may be due to significant interfacial friction between the fiber and soil matrix with confinement bonding (Anagnostopoulos *et al.* 2014). Figure 9b shows the stress-strain curve of BCS stabilized with different cementitious binders and fibers. Raw BCS attains a low peak strength at around 2.3% of strain and PP fiber reinforced with AAB treated BCS attains highest shear strength with the lowest strain. The drastic improvement in compressive strength can be because of geopolymeric reaction induced between the sodium aluminosilicate and pozzolanic additives in the clay particles (Malik and Priyadarshee 2018).

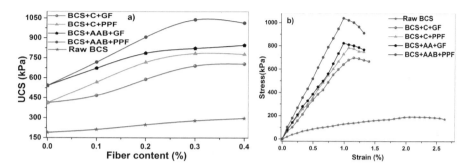

Fig. 9. Variation of UCS (a) Stress-strain curves of untreated BCS and fiber reinforced cement and AAB treated BCS

Indirect Tensile Strength (ITS)

Indirect tensile tests are conducted as per ASTM D4123-1995 standard on cement and AAB treated soil incorporated with different fibers. The soil specimens are prepared by maintaining 80 mm height and 100 mm diameter with the loading strip of 12.5 mm at a constant strain rate of 50.5 mm/min in a Marshall stability machine. The samples are preserved for 24 h in the humidity chamber before testing (Emesiobi 2001; Eberemu and Sada 2013). Figure 10a shows the variations of ITS values of untreated BCS, as well as fiber, reinforced cement, and AAB, treated BCS. It can be observed that the tensile strength of cement and AAB treated BCS increases with increase in glass and polypropylene fiber content. Upon comparison, it can be seen that the AAB treated BCS reinforced with polypropylene fiber achieves the highest tensile strength. This drastic improvement in ductile properties and stretching resistance is majorly due to fiber surface morphology and pozzolanic reaction (Tang *et al.* 2007; Moghal *et al.* 2017). Figure 10b shows the typical arrangement of indirect tensile soil strength specimen under the loading strip frame. Hence the fiber-AAB compound mixture can effectively reduce the brittleness behavior and control the formation of the tensile cracks.

California Bearing Ratio (CBR)

Soaked and unsoaked CBR tests are performed for both treated and untreated BCS using ASTM D-1883 standard. Figure 11 shows the variations of CBR results for both soaked and unsoaked values at 2.5 mm penetration. As seen from the graph, the soaked CBR value of raw BCS is 1.96, indicating low strength and bearing. It is also noticed that the CBR value of cement and AAB treated soil increases with an increase in fiber content. In addition, the highest soaked and unsoaked CBR values are found at 0.3%

polypropylene fiber with 100% fly ash based AAB treated BCS when compared to the fiber-cement mixture. Blending of cementitious and pozzolanic compounds with fiber in the BCS, aid to upgrade the strength bearing ratio of soil through hydration and geopolymerisation reaction during the period of soaking (Das 2003; Kumar *et al.* 2007; Malik and Priyadarshee 2018).

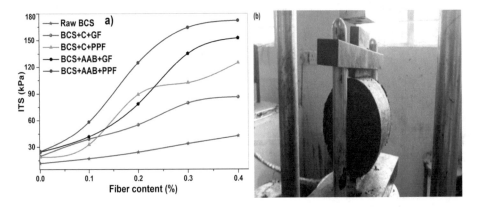

Fig. 10. Variation of ITS of untreated BCS and fiber reinforced cement and AAB treated BCS (a) Typical arrangement of ITS test

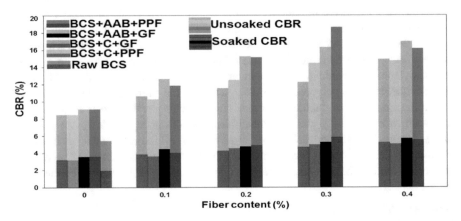

Fig. 11. Variation of soaked and unsoaked CBR values of untreated BCS and fiber reinforced cement and AAB treated BCS

4   Summary and Conclusions

The present study compares the microstructure and geoengineering characteristics of glass and polypropylene fiber based soil reinforcement with a traditional cement binder and envirosafe alkali-activated binders (AAB). Fly ash based geopolymer binder serves a dual benefit of preventing the disposal of industrial by-product and at the same time maintaining a green sustainable environment. The important findings and main conclusions can be summarized as follows.

- Microanalysis results confirm the formation of new crystalline phases and molecular vibrations in the cement and AAB treated BCS. In addition, stereomicroscopic images of glass and polypropylene fiber reinforced with cement or AAB show improved bonding behavior.
- Micrographs of fiber-cement and fiber-AAB treated BCS shows strong interfacial surface interaction between the fiber matrix and soil through the formation of cementitious structure in the soil mass.
- The geotechnical results show that AAB treated BCS with polypropylene fiber shows the most significant improvement in tensile, bearing capacity and shear strength properties of BCS when compared to glass fiber- AAB treated and cement-fiber treated BCS.
- Void ratio and swelling pressure of cement treated BCS is significantly reduced by 36%, whereas AAB treated BCS reduces the same by about 32%. The volumetric stability and tensile resisting properties of BCS is greatly enhanced by blending 0.3% of polypropylene fiber with AAB.
- Strength bearing ratio in terms of CBR, compaction, and shear strength values of glass and polypropylene fiber reinforced AAB treated BCS increases by 46% and 58%, whereas cement treated BCS increases by 38% and 45% respectively.

Acknowledgements. The authors would like to express their sincere gratitude to the Central Analytical Laboratory Facilities at BITS-Pilani, Hyderabad Campus for providing the setup for the XRD, FTIR, and SEM-EDS analyses.

References

Ackroyd, L.W., Husain, R.: Residual and lacustrine black cotton soils of north-east Nigeria. Geotechnique **36**(1), 113–118 (1986)

Al-Rawas, A.A., Hago, A.W., Al-Sarmi, H.: Effect of lime, cement and Sarooj (artificial pozzolan) on the swelling potential of an expansive soil from Oman. J. Build. Environ. **40**, 681–687 (2005)

Anagnostopoulos, C.A., Tzetzis, D., Berketis, K.: Shear strength behaviour of polypropylene fibre reinforced cohesive soils. Geomech. Geoeng. **9**(3), 241–251 (2014)

Bell, F.G.: Lime stabilization of clay minerals and soils. Eng. Geol. **42**(4), 223–237 (1996)

Chen, F.H.: Foundations on Expansive Soils. Elsevier Scientific Publishing Co., Amsterdam (1988)

Chen, H., Wang, Q.: The behaviour of organic matter in the process of soft soil stabilization using cement. Bull. Eng. Geol. Environ. **65**(4), 445–448 (2006)

Das, B.M.: Chemical and mechanical stabilization. Transportation Research Board (2003)

Davidovits, J.: Properties of geopolymer cements. In: First International Conference on Alkaline Cements and Concretes, Kiev State Technical University, Scientific Research Institute on Binders and Materials, Ukraine, vol. 1, p. 13 (1994)

Eberemu, A.O., Sada, H.: Compressibility characteristics of compacted black cotton soil treated with rice husk ash. Niger. J. Technol. **32**(3), 507–521 (2013)

Gartner, E.: Industrially interesting approaches to 'low-CO_2' cements. Cem. Concr. Res. **34**(9), 1489–1498 (2004)

Gupta, S., GuhaRay, A., Kar, A., Komaravolu, V. P.: Performance of alkali-activated binder-treated jute geotextile as reinforcement for subgrade stabilization. Int. J. Geotech. Eng. 1–15 (2018)

Kaniraj, S.R., Gayathri, V.: Geotechnical behavior of fly ash mixed with randomly oriented fiber inclusions. Geotext. Geomembr. **21**, 123–149 (2003)

Katti, R.K.: Search for solutions to problems in black cotton soils. Indian Institute of Technology, Bombay (1978)

Kumar, A., Walia, B.S., Bajaj, A.: Influence of fly ash, lime, and polyester fibers on compaction and strength properties of expansive soil. J. Mater. Civ. Eng. **19**(3), 242–248 (2007)

Lin, B., Cerato, A.B., Madden, A.S., Elwood Madden, M.E.: Effect of fly ash on the behavior of expansive soils: microscopic analysis. Environ. Eng. Geosci. **19**(1), 85–94 (2013)

Madejova, J., Komadel, P.: Baseline studies of the clay minerals society source clays: infrared methods. Clays Clay Miner. **49**(5), 410–432 (2001)

Malik, V., Priyadarshee, A.: Compaction and swelling behavior of black cotton soil mixed with different non-cementitious materials. Int. J. Geotech. Eng. **12**(4), 413–419 (2018)

Mazhar, S., GuhaRay, A., Kar, A., Avinash, G.S.S., Sirupa, R.: Stabilization of expansive black cotton soils with alkali activated binders. In: Proceedings of China-Europe Conference on Geotechnical Engineering, pp. 826–829. Springer, Cham (2018)

Moghal, A.A.B., Chittoori, B., Basha, B.M., Al-Shamrani, M.A.: Target reliability approach to study the effect of fiber reinforcement on UCS behavior of lime treated semiarid soil. J. Mater. Civ. Eng. **29**, 04017014 (2017). https://doi.org/10.1061/(ASCE)MT.1943-5533.0001835

Miao, S., Shen, Z., Wang, X., Luo, F., Huang, X., Wei, C.: Stabilization of highly expansive black cotton soils by means of geopolymerisation. J. Mater. Civ. Eng. **29**(10), 04017170 (2017)

Ural, N.: Effects of additives on the microstructure of clay. J. Road Mater. Pavement Des. **10**, 1–16 (2015)

Ola, S.A.: Geotechnical properties and behavior of some stabilized Nigerian lateritic soils. Q. J. Eng. Geol. Hydrogeol. **11**(2), 145–160 (1978)

Oren, A.H.: Estimating compaction parameters of clayey soils from sediment volume test. Appl. Clay Sci. **101**, 68–72 (2014)

Ouhadi, V.R., Yong, N.R.: Ettringite formation and behaviour in clayey soils. Appl. Clay Sci. **42**, 258–265 (2008)

Phani Kumar, B.R., Sharma, R.S.: Effect of fly ash on engineering properties of expansive soils. J. Geotech. Geoenviron. Eng. **130**(7), 764–767 (2004)

Petry, T.M., Little, D.N.: Review of stabilization of clays and expansive soils in pavements and lightly loaded structures—history, practice, and future. J. Mater. Civ. Eng. **14**(6), 447–460 (2002)

Rios, S., Cristelo, N., Viana da Fonseca, A., Ferreira, C.: Structural performance of alkali-activated soil ash versus soil cement. J. Mater. Civ. Eng. **28**(2), 04015125 (2015)

Saride, S., Dutta, T.T.: Effect of fly-ash stabilization on stiffness modulus degradation of expansive clays. J. Mater. Civ. Eng. **28**(12), 04016166 (2016)

Salahudeen, A.B., Eberemu, A.O., Osinubi, K.J.: Assessment of cement kiln dust-treated expansive soil for the construction of flexible pavements. Geotech. Geol. Eng. **32**(4), 923–931 (2014)

Sharma, N.K., Swain, S.K., Sahoo, U.C.: Stabilization of a clayey soil with fly ash and lime: a micro level investigation. Geotech. Geol. Eng. **30**(5), 1197–1205 (2012)

Sekhar, D., Nayak, S.: SEM and XRD investigations on lithomargic clay stabilized using granulated blast furnace slag and cement. Int. J. Geotech. Eng. **13**, 1–15 (2017)

Sivapullaiah, P.V., Prasad, B.G., Allam, M.M.: Effect of sulfuric acid on swelling behavior of an expansive soil. Soil Sediment Contam. **18**(2), 121–135 (2009)

Tang, C., Shi, B., Gao, W., Chen, F., Cai, Y.: Strength and mechanical behavior of short polypropylene fiber reinforced and cement stabilized clayey soil. Geotext. Geomembr. **25**(3), 194–202 (2007)

Vitale, E., Russo, G., Dell'Agli, G., Ferone, C., Bartolomeo, C.: Mechanical behaviour of soil improved by alkali activated binders. Environments **4**(4), 80 (2017)

Yetimoglu, T., Inanir, M., Inanir, O.E.: A study on bearing capacity of randomly distributed fiber-reinforced sand fills overlying soft clay. Geotext. Geomembr. **23**(2), 174–183 (2005)

Zhao, H., Ge, L., Petry, T., Sun, Y.-Z.: Effects of chemical stabilizers on an expansive clay. J. Civ. Eng. KSCE **10**, 1–9 (2013)

On the Consequences of Microstructural Evolution on Macroscopic Behavior for Unsaturated Soils

Hiram Arroyo[1($\boxtimes$)], Eduardo Rojas[2], Jatziri Y. Moreno-Martínez[1], and Arturo Galván[1]

[1] Universidad de Guanajuato, Celaya, Mexico
hiramarroyo@gmail.com
[2] Universidad Autónoma de Querétaro, Querétaro, Mexico

Abstract. Volumetric behavior of soils is amongst the most important parameters to be taken into consideration to describe many aspects regarding the hydro-mechanical coupling of unsaturated soils, such as the behavior of CO_2 reservoirs subjected to moisture gradients, or the initiation of cracks due to suction loading-reloading.

Experimental observations show a clear transition of compressibilities as soils reach unsaturated states when tested under suction increments. This is termed shrinking limit. We address this behavior taking into consideration the evolution of pores where a mechanism of pore deformation is proposed to predict this shrinking limit. This model requires no other parameter than the initial pore-size distribution of the material.

1 Introduction

It is believed that the explanation for every observation regarding the mechanical behavior of soils lies on clarifying the mechanism of how pores behave on the presence of the liquids that they appertain within their pores.

When it comes to volumetric strains, experimental observations indicate that the compressibility due to suction increments highly decrease on the transition to saturated states. This transition has been identified by Perón (2008) as the shrinking limit where different compressibilities are needed to describe macroscopic evolution of volume strains. Constitutive models take this into consideration adding parameters to identify this transition, however, this increases the number of needed tests to calibrate such models.

Authors believe that conceptualizing the evolution of the pore-size distribution is the key to understand and predict macroscopic evidences such as the shrinking limit. Experimental observations evidence that pore-size distribution of soils exhibit most of the times a bimodal distribution (Alonso et al. 2010; Cui et al. 2002; Koliji et al. 2006). This pore-size distribution is related to macroscopic observations, as has been reported that macro-pores are the ones responsible for most of volume change observed. There is on the other hand a micro-pore distribution that suffers negligible changes as suction is applied. Here, suction influence on volumetric behavior is predominant when the soil is in a saturated state (Perón et al. 2009).

© Springer Nature Switzerland AG 2020
L. Hoyos and H. Shehata (Eds.): GeoMEast 2019, SUCI, pp. 56–66, 2020.
https://doi.org/10.1007/978-3-030-34206-7_5

This paper aims at proposing a mechanism for pore-size distribution evolution with suction increments to predict the shrinking limit of materials and the evolution of water retention properties of soils.

2 The Squared Grid Model

Because the internal structure of soils is too complicated to be modeled, the pore structure is simplified to regular geometries. It has been shown that the porous network contained within solid particles, can be modeled considering two main types of entities: sites and bonds. Sites contribute with most of the void space available for deformation and are responsible for every macroscopical aspect of soils that involves volumetric strains. Bonds interconnect sites and have little contribution to the volume of voids (Fig. 1).

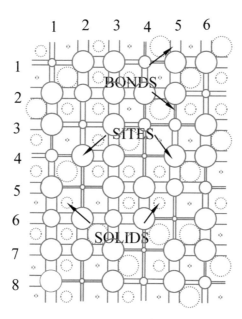

Fig. 1. Porous network used to model the porous space of soils (Arroyo and Rojas 2019).

The grid is composed of a set of interconnected sites. The connection elements between sites are called bonds. One solid particle is contained within a set of four sites interconnected by four bonds.

3 The Solid-Porous Model by Rojas

3.1 Modelling Pore-Size and Grain-Size Distributions

A first approach is to consider that all soil particles appertain a regular circular shape for the case of a 2D problem (spherical when it comes to dealing with 3D situations).

In order to make the model as real as possible, the frequency $f = n/N$ needs to be established. The frequency f with which each class of grains (i.e. set of n solid grains having the same size) appears within a soil sample, is referred to its total number of grains N. Here, to model f, a normal logarithmic density function is used:

$$f(R) = \frac{1}{\sigma R \sqrt{2\pi}} \exp\left[\frac{-(\ln R - \mu)^2}{2\sigma^2}\right] \tag{1}$$

Where, $f(R)$ is the frequency distribution of the class of grains of size R contained within the soil sample. Considering the network of solid grains, the volume of grains of size R is given by:

$$V^{SOL}(R) = N\pi \int_{R}^{R+dR} f(R)R^2 dR \tag{2}$$

Moreover, because soils can appertain clayey size particles, as well as sand particles, two functions can be used. Therefore, Eq. (2) can be expressed as $V^{SOL}(R) = N\pi \int_{R}^{R+dR} \left(f^{MSOL}(R)R^2 + f^{mSOL}(R)R^2\right)dR$. An additional degree of freedom is provided to model the GSD. This is, the inclusion of factor F_{SOL}^p, which is necessary to include due to the logarithmic scales of Eq. (1). This factor reduces the height of f^{MSOL}. Therefore, Eq. (2) can be expressed as:

$$V^{SOL}(R) = N\pi \int_{R}^{R+dR} \left(F_{SOL}^p f^{MSOL}(R)R^2 + f^{mSOL}(R)R^2\right)dR \tag{3}$$

The process to model the Pore-Size Distribution (PSD) is analogous to the one followed for the GSD. The following equation is used for the PSD:

$$V^{S}(R) = N\pi \int_{R}^{R+dR} \left(F_{S}^p f^{MS}(R)R^2 + f^{mS}(R)R^2\right)dR \tag{4}$$

Functions f^{MS} and f^{mS} are used to model the PSD of materials that appertain two classes of pores (large and small). Because soils are highly heterogeneous materials, the individual solids they contain have very different ranges of sizes. This double structure is produced by the compaction process and the initial way in which the soil sample was set up in the testing device. To differentiate two classes of sites, the terms "macro-sites" and "micro-sites" will be used regardless other classifications.

Contrary to the case of f^{MSOL} and f^{mSOL}, parameters controlling PSD functions are not constant. This implies that solid phase is incompressible, the void phase is not. The behavior of these two functions is the key of this work and it is believed that every single aspect of volumetric behavior (saturated and unsaturated), and the way to conceal saturated and unsaturated soil mechanics at the highest level using a simple constitutive model, such as the Cam Clay (Roscoe and Burland 1968), can be done by correctly modeling their behavior.

A pore-size distribution for bonds is also required, since the size of the interconnections between site elements depend on the size of the pores they interconnect.

$$V^B(R) = N\pi \int_R^{R+dR} \left(F_B^p f^{MB}(R)R + f^{mB}(R)R \right) dR \tag{5}$$

3.2 GSD and PSD Models to Determine the Void Ratio of the Material

Integrating both, Eqs. (3) and (4), over the whole range of possible sizes of particles allow determining the total volume of solid phase $\overline{V^{SOL}}$ and void phase $\overline{V^S}$. Therefore, void ratio can be expressed as:

$$e = \frac{\overline{V^S}}{\overline{V^{SOL}}} = \frac{N_S \pi \int_0^{inf} \left(F_S^p f^{MS}(R)R^2 + f^{mS}(R)R^2 \right) dR}{N_{SOL} \pi \int_0^{inf} \left(F_{SOL}^p f^{MSOL}(R)R^2 + f^{mSOL}(R)R^2 \right) dR} = \frac{N_S \overline{V^{S^*}}}{N_{SOL} \overline{V^{SOL^*}}} \tag{6}$$

3.3 Determination of Parameters for the GSD and PSD

Parameters to evaluate Eq. (3) describing material GSD are μ_{SOL}^M, σ_{SOL}^M, μ_{SOL}^m and σ_{SOL}^m. This is, a set of μ and σ parameters for both f^{MSOL} and f^{mSOL}. These are determined by a fitting procedure where the goal is to graphically superimpose the differential GSD of the material. The real GSD is obtained by any experimental means. Figure 2 is an example of the usage of Eq. (3).

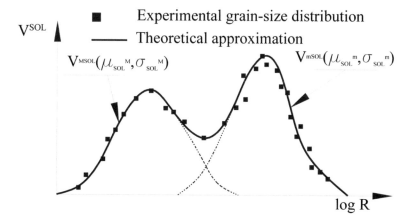

Fig. 2. Fitting theoretical to experimental data of solids volume distribution.

Once fitted, the total volume of solid particles the soil sample appertains, can be computed integrating Eq. (3) from the smallest particle to the largest one:

$$\overline{V^{SOL}} = N\pi \int_0^{inf} \left(F_{SOL}^p f^{MSOL}(R)R^2 + f^{mSOL}(R)R^2 \right) dR \tag{7}$$

Obtaining parameters to determine the PSD (Eq. (4)) is not as straightforward. To do this, the fact that there is a unique one to one relationship between a material's PSD and its water retention curve (Dullien 1992; Haines 1927), will be used. This is, parameters for Eq. (4) are those that will serve to fit the experimental Water Retention Curve (WRC).

This process begins constructing a rectangular grid. For this, a set of parameters to evaluate f^{MSOL} and f^{mSOL} are proposed, these will serve to evaluate Eq. (4) which will be a first approach to the PSD.

Once the tentative PSD is evaluated, the number n of sites required to fulfill the needs of the grid can be determined since $f = n/N$. It is important to notice that, randomness is of paramount importance. Therefore, the grid will be constructed taking an element corresponding to each size of the grid randomly and placing it randomly within the grid until very spot of the network is filled by a site. The same happens for the bonds distribution.

During a wetting process, all sites are initially dried, then suction is reduced by steps, and the grid is analyzed to identify those cavities that can be filled with water (see Fig. 3(a)). As stated by Rojas et al. (2011) This is done by identifying those cavities that comply with the following rules that lead pores to saturate: (a) its radius R must be smaller than the critical value R_C given by the Young–Laplace equation, and (b) it must be connected to the bulk of water following a continuous path from the boundaries.

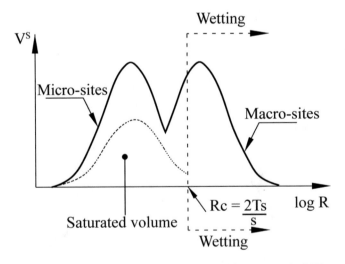

Fig. 3. Wetting-drying processes simulated from a certain PSD.

The Young-Laplace equation writes $s = \mu_a - \mu_w = 2T_S \cos \phi / R_C$ (Lu and Likos 2004), where ϕ is the contact angle which is usually considered as 0 and $T_S = 7.36 \, \text{N/m}$ represents the air-water interfacial tensional force. Hence the critical radius can be computed as follows:

$$R_C = \frac{2T_S}{S} \tag{8}$$

For the case of a drying process, a site will dry when it complies with the following conditions: (a) the radius R of the site is larger than the critical radius R_C and (b) it must be connected to the boundaries by a continuous path of dried pores.

With every suction increase/decrease, the rectangular grid is analyzed to search for sites that comply with these two conditions during a drying or wetting path, respectively. It can be seen from Fig. 3 that the total volume supersedes the saturated volume of pores. This is because not all sites are able to saturate or dry according to the previous two established conditions.

The water retention expression is then, that of:

$$Sr(R) = \frac{Vw}{Vv} \tag{9}$$

Where Vw is the volume of water-saturated sites that the grid appertains up to the suction value corresponding to R_C, and Vv is the total volume of the sites of the grid.

3.4 Pore-Size Distribution Evolution

Several reported results on the evolution of the PSD with volumetric strains induced either by increasing suction or by applied mean net stresses indicate that the macropores distribution is the one that seems to be affected the most (Cui et al. 2002; Koliji et al. 2006; Simms and Yanful 2005; Thom et al. 2007). That is, function $f^{MS}\left(\mu_{SOL}^M, \sigma_{SOL}^M\right)$ changes by shifting towards smaller sizes as the soil reduces its void ratio by any means. However, its shape does not seem to change. For this reason, compressive strains will be expressed as a shifting of the f^{MS} changing μ_{SOL}^M solely.

4 Numerical-Experimental Result Comparisons

Perón (2008) evaluate the water retention curve of a Bioley Silt, an inorganic clay of medium plasticity. Samples were fabricated with a water content of 1.5 times their liquid limit, thus ensuring a saturated state. Then, the wet material was poured into a small cylindrical mold of 50 mm diameter and 10 mm height. A goal of this was to study the volume and saturation changes of soils induced by suction The molds containing the wet samples were placed within a pressure plate extractor to measure the water retention curve.

Figure 4 shows the cumulative theoretical GSD which was fitted to the experimental GSD using Eq. (3). Figure 5 shows a theoretical-experimental fitting of the soil-water characteristic curve using the porous model herein.

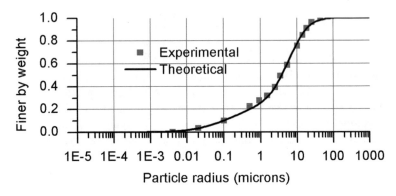

Fig. 4. GSD (Fitted and experimental) $\mu_{SOL}^M = -0.2$, $\sigma_{SOL}^M = 1.0$, $\mu_{SOL}^m = -9.5$, $\sigma_{SOL}^m = 2.0$, and $F_{SOL}^p = 0.000008$.

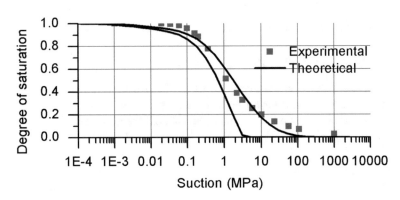

Fig. 5. Differential PSD (fitted and experimental) $\mu_S^M = -2.5$, $\sigma_S^M = 1.7$, $\mu_S^m = -8.05$, $\sigma_S^m = 1.7$ and $F_S^p = 0.000001$.

The resulting PSDs and GSDs (predicted) are depicted in Fig. 6. Very important to note from Fig. 6 is that superimposing the macro-solids and macro-sites distributions exhibits a clear resemblance on the range of sizes distribution (peak and distribution). The same for the micro-solids and the micro-sites distributions. This may be an intuitive idea, however, the model confirms this, as stated by Rojas et al. (2011).

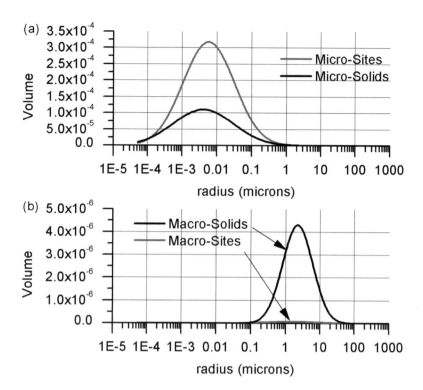

Fig. 6. (a): Predicted micro-solids and micro-sites distributions. (b): Predicted macro-solids and macro-sites distributions.

The relationship between void ratio and suction experimentally evaluated by Perón (2008) is contained in Fig. 7. The initial void ratio of the saturated poured slurry is 0.82 at suction of 19 kPa. The shrinking limit is at a void ratio between 0.60 and 0.58 and a suction value that ranges between 131 and 196 kPa. Using Eq. (6) it is possible to

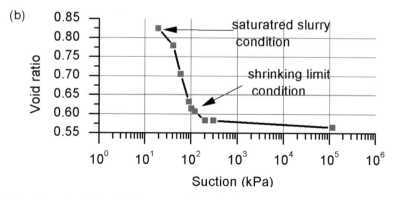

Fig. 7. Experimental relationship between void ratio and suction increments for a Bioley Silt (Perón 2008).

compute the void ratio of the material corresponding to the PSD and GSD of Fig. 6. This retrieves a void ratio of 0.62 which resembles to the real shrinking limit condition.

The model is used to predict the evolution of the PSD. Figure 8 depicts the macropore evolution at the initial saturated slurry and at the shrinking limit condition. This was achieved by increasing the μ_{SOL}^{M} parameter solely as explained earlier.

The macropores depicted on Fig. 8 are used to compute the WRCs at the initial saturated slurry conditions and at the shrinking limit (see Fig. 9).

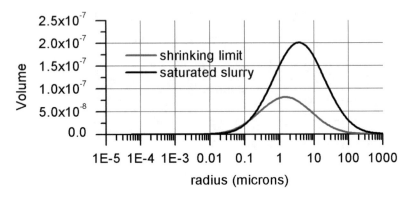

Fig. 8. Predicted evolution of macro-sites to represent the WRC evolution.

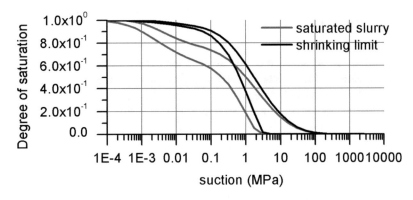

Fig. 9. Predicted WRC at the initial conditions and at the shrinking limit.

Note that, as stated by Perón (2008), the shrinking limit clearly appears at the transition between the saturated and unsaturated condition (i.e. at the suction corresponding to the air entry value).

Here, the shifting of the WRC depicted in Fig. 9, predicted with the presented solid porous model, is in accordance with experimental observations where different WRCs are evaluated for soils with different densities (Ng and Pang 2000; Sun et al. 2007). This shifting occurs due to the compression of the larger pores at early stages of suction increments which make all the voids remain in a saturated state regardless of the material water content. Further suction increments seem to have a reduced amount of importance on volume strains after the air entry value as it is for the shrinking limit of the material. This is of paramount importance, since this leads to additional tensile stresses that are the reactions to the tendency of the water in tensile state to shrink the material, leading to the to tensile strength as reported by Perón (2008).

5 Conclusions

A mechanism for the evolution of the PSD has been proposed. This enables coupling a constitutive elastoplastic model at high level due to the natural coupling between macroscopic strains and microstructure provided by the solid porous model presented in this paper. This arises from a comprehensive understanding of the behavior and modeling of microstructure considering the PSD as its main expression. Further work needs to be done to understand the evolution of PSD for highly expansive soils.

Acknowledgments. Financial support from the Universidad de Guanajuato in Mexico is greatly acknowledged.

References

Alonso, E.E., Pereira, J.M., Vaunat, J., Olivella, S.: A microstructurally based effective stress for unsaturated soils. Géottechnique **60**(12), 913–925 (2010)

Arroyo, H., Rojas, E.: Fully coupled hydromechanical model for compacted soils. Comptes Rendus Mecanique **347**(1), 1–18 (2019)

Cui, Y.J., Loiseau, C., Delage, P.: Microstructure changes of a confined swelling soil due to suction controlled hydration. In: Jucá, J.F.T., Campos, T., Marinho, F.A.M. (eds.) Unsaturated Soils, pp. 593–598. Balkema, Leiden (2002)

Dullien, F.A.L.: Porous Media: Fluid Transport and Pore Structure. Academic Press, San Diego (1992)

Haines, W.B.: Studies in the physical properties of soils: IV. A further contribution to the theory of capillary phenomena in soil. J. Agric. Sci. **17**(02), 264–290 (1927)

Koliji, A., Laloui, L., Cuisinier, O., Vulliet, L.: Suction induced effects on the fabric of a structured soil. Transp. Porous Med. **64**, 261–278 (2006)

Lu, N., Likos, W.J.: Unsaturated Soil Mechanics. Wiley, New York (2004)

Ng, C.W.W., Pang, Y.W.: Influence of stress state on soil-water characteristics and slope stability. J. Geotech. Geoenviron. Eng. **126**, 157–166 (2000)

Perón, H.: Desiccation Cracking of Soils (Ph.D.), École Polytechnique Fédérale de Lausanne (2008)

Perón, H., Hueckel, T., Laloui, L., Hu, L.: Fundamentals of desiccation cracking of fine-grained soils: experimental characterisation and mechanisms identification. Can. Geotech. J. **46**(10), 1177–1201 (2009)

Rojas, E., Horta, J., López-Lara, T., Hernández, J.B.: A probabilistic solid porous model to determine the shear strength of unsaturated soils. Prob. Eng. Mech. **26**(3), 481–491 (2011)

Roscoe, K.H., Burland, J.B.: On the generalized stress-strain behavior of 'wet' clay. Engineering Plasticity, pp. 535–609. Cambridge University Press, Cambridge (1968)

Simms, P.H., Yanful, E.K.: A pore-network model for hydro-mechanical coupling in unsaturated compacted clayey soils. Can. Geotech. J. **42**(2), 499–514 (2005)

Sun, D.A., Sheng, D.C., Cui, H.B., Sloan, S.W.: A density-dependent elastoplastic hydro-mechanical model for unsaturated compacted soils. Int. J. Numer. Anal. Meth. Geomech. **31** (11), 1257–1279 (2007)

Thom, R., Sivakumar, R., Sivakumar, V., Murray, E.J., Mackinnon, P.: Pore size distribution of unsaturated compacted kaolin: the initial states and final states following saturation. Géotechnique **57**(5), 469–474 (2007)

Investigation on Geotechnical Properties Before and After the Construction of Earth Retaining Structures-West Konkan a Case Study

Arun Dhawale[1], Sudarshan Sampatrao Bobade[2], Anand Tapase[3(✉)],
and Vaibhav Garg[4]

[1] Department of Civil Engineering, Imperial College of Engineering
and Research, Pune, Maharashtra, India
awdhawale2009@gmail.com
[2] Department of Civil Engineering, Bhivrabai Sawant College of Engineering
and Research, Pune, India
bsudarshan8376@gmail.com
[3] Department of Civil Engineering, Rayat Shikshan Sanstha's Karmaveer
Bhaurao Patil College of Engineering, Satara, Maharashtra, India
tapaseanand@gmail.com
[4] Water Resource Department, Indian Institute of Remote Sensing (ISRO),
Dehradun, India
vaibhav@iirs.gov.in

Abstract. Characteristic geotechnical properties of soil vary from place to place. The classification of soil mass depends on rock family from which it originates its mineral composition and the environmental aspect of the area. Civil engineering structure is founded in or on the surface of the earth, and hence before designing any structures, it is necessary to closely observe the suitability of a soil for construction. Properties of soils observed before construction and after the construction of earth retaining structures (ERS) influences the stability of structures. Most of the geotechnical properties of surrounding soils in the vicinity of ERS changes after the construction of the ERS. In a developing country like India, the population density is very high due to which people are forced to reside in landslide-prone areas and this initiates the need to construct Earth Retaining Structures. The stability of these ERS is very crucial to protect the vulnerable slopes. If the ERS are displaced even slightly it can create havoc by bringing the lives and property in danger. In this paper, different geotechnical properties of soils such as specific gravity, density index, consistency limits, particle size analysis, compaction, consolidation, permeability, and shear strength have been studied for observing the changes in geotechnical properties of the soil before and after the construction of Earth Retaining Structures and conclusions are noted. This paper attempts to find out the correlation between the displacement of ERS against changes in soil characteristics using GNSS technology. It is observed from past research studies that GNSS technology can be used for tracing out the precise location of any object or structure. Keeping this in mind, the soil sample in the vicinity of ERS was regularly checked in the laboratory to correlate the soil characteristics and

© Springer Nature Switzerland AG 2020
L. Hoyos and H. Shehata (Eds.): GeoMEast 2019, SUCI, pp. 67–80, 2020.
https://doi.org/10.1007/978-3-030-34206-7_6

location of the ERS. The practice of testing the soil characteristics in the laboratory was consistently followed. From the laboratory and field tests observation, it has been observed that soil characteristics of the soil in the vicinity of ERS change even with the minor displacement of the ERS from its initial position. Hence, GNSS technology can be used to give early warnings related to major displacements of ERS that could take place in the future. Thus GNSS technology can be used to produce a low-cost early warning system for the displacement of ERS so that a warning about a probable landslide can be received well in advance to evacuate the area and save life and property.

Keywords: Geotechnical properties · Earth retaining structures · Drainage · Shear strength-bearing capacity of soil · GNSS

1   Introduction

Landslide Hazard Mitigation as disaster management is a very tough job for a Civil, Structural, and Geotechnical Engineer. In hilly terrain, due to manmade and natural activities landslide occurs frequently. It affects the lifestyle of local people, ecosystem, and economy of that area. The soil stabilization techniques are introduced to reduce such landslide; the earth retaining structures provision is one of them.

Earth Retaining Structures (ERS) are relatively rigid walls used for supporting the soil mass laterally so that the soil can be retained at different levels on the two sides. They are used to bound soils between two different elevations often in areas of terrain possessing undesirable slopes or in areas where the landscape needs to be shaped severely and engineered for more specific purposes like hillside farming or roadway overpasses.

The geotechnical, structural and economic consideration is the most important to design the earth retaining structure. To design earth retaining structure it is important to know the characteristic geotechnical properties of soil in that area, which effects on the serviceability of that ERS. Properties of soils observed before construction and after the construction of earth retaining structures (ERS) influences the stability of structures. To study this changes various geotechnical tests are conducted on backfill material against precise positioning study of ERS using GNSS and its impact on design serviceable life in terms of factor of safety is calculated.

Majorly retaining structure was failed in the rainy season; therefore it triggers need to compute changes in input design parameters at current condition and at the full saturated condition of the soil. In the case of ERS, its structural design is mainly dependent on cohesion, internal angle of friction, soil bearing capacity, the density of soil, surcharge angle of backfill and height of retaining wall. Among these listed parameters cohesion, internal angle of friction, dry density and saturated density of soil are found to be dependent on % of fine aggregate and Natural moisture content in the composition of backfill material. As percentage fineness is increased it affects on the specific gravity, natural moisture content, cohesion, dry density and saturated density of soil.

This paper attempts to find out the correlation between the displacements of ERS using GNSS technology against changes in soil characteristics if any. It is observed from past research studies that GNSS technology can be used for tracing out the precise location of any object or structure. Keeping this in mind, the soil sample in the vicinity of ERS was regularly checked in the laboratory to correlate the soil characteristics and location of the ERS. The practice of testing the soil characteristics in the laboratory was consistently followed. From the laboratory and field tests observation, it has been observed that soil characteristics of the soil in the vicinity of ERS change even with the minor displacement of the ERS from its initial position. Hence, GNSS technology can be used to give early warnings related to major displacements of ERS that could take place in the future. Thus GNSS technology can be used to produce a low-cost early warning system for the displacement of ERS so that a warning about a probable landslide can be received well in advance to evacuate the area and save life and property.

2 Study Area and Geology

Detailed investigation of a major landslide against relative mitigative measures for its serviceability at Dasgaon in South – Western Maharashtra i.e. Kokan region has been presented in this paper. Location for this cantilever ERS is in Raigad District having Latitude- N18° 6′ 46.08″ and Longitude- E 73° 21′ 54″ Eat Dasgaon. The total height of Dasgaon cantilever ERS is 7.5 m. Geologically this region has Deccan Trap Basaltic rocks and lateritic rock, which are weathered near the surface due to highly oxidizing and humid climatic conditions, developing lithomarge clays and lateritic soil regolith. Loose cohesive soil matrix becomes soft and loses strength due to surface and sub-surface flow during heavy precipitation. Regolithic mass of overburden became heavy, lost support/interlocking and slumped along the hill slope. This region lies in zone IV as per map of seismic zones for our country. Therefore minute seismic event may act as a triggering factor leading to slope instability (Map 1).

The hilly terrain of Dasgaon area is south-western part of Sahyadri in the coastal strip. It is composed of the oldest rock formation of the world that is Archaean systems. This rock formation after weathering has converted to the relict hilly terrain with lateritic soil mass resting on the angle of repose of gneisses. This hilly terrain in coastal zone experiences very heavy rainfall during monsoon. It is seismically active and comes under zone IV as per seismic zonation map. Therefore Landslide is a frequently observed phenomenon in this area during monsoon.

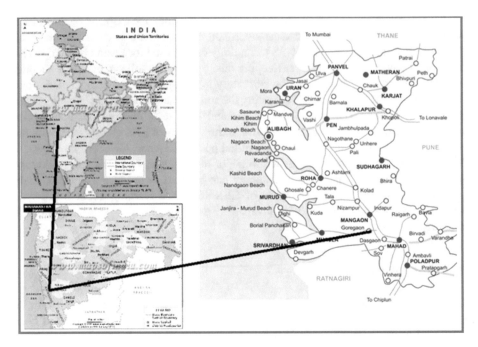

Map 1. Study Area

3 Study of Characteristic Geotechnical Properties of Soil

3.1 Input Design Parameters at the Time of Structural Design

The Table 1 shows the input design parameters at the time of structural design for Dasgaon cantilever ERS. The properties of concrete and steel are M 30 and Fe500. Height of retaining structure is 7.5 m for Dasgaon.

Table 1. Input design parameters at the time of structural design

Sr. No.	Parameters	Pre values
1	Specific gravity	2.9
2	The dry density of soil (KN/m^3)	18
3	The saturated density of soil (KN/m^3)	20
4	The cohesion of soil (KN/m^2)	0
5	Internal angle of friction (°)	30

3.2   Input Design Parameters at the Current Scenario

To analyze the condition of backfill material at the current state the following tests were conducted on backfill material. Soil samples are collected from the backfill of the dasgaon retaining structure at two conditions from three different locations. The reasons to check these tests is that this design parameter effects on the Factor of safety considered while design.

1. The determination of natural moisture content of soil by oven drying method.
2. The determination of specific gravity of soil.
3. Sieve analysis by mechanical method.
4. Determination of Atterberg's Limit.
5. Determination of Maximum dry density and Optimum Moisture Content of soil by Standard Proctor test.
6. Determination of cohesion and internal angle of friction by the direct shear method.

Tables 2 and 3 shows the input design parameters at the current condition by conducting the laboratory tests on a borrowed soil sample. This input design parameter effects on the serviceability in terms of calculation of factor of safety, therefore it is essential to find out the input design parameter.

Table 2. Characteristics Properties of soil at Dasgaon-Position 1 through GNSS - Latitude-N18° 6′ 46.08″ and Longitude- E 73° 21′ 54″

Sr. No	Parameters	Sample A	Sample B	Sample C
1	Natural moisture content	16.95	11.39	17
2	Specific gravity	2.48	2.56	2.54
3	% of soil passing through 200 No. Sieve	18.6	24.3	17.9
4	Liquid limit	48.39	41.63	43.78
5	Plastic limit	36.43	33.26	37.56
6	Plasticity index	11.96	8.37	6.22
7	Optimum moisture content	1.98	1.43	1.394
8	Maximum dry density	25.05	28.47	34
9	The dry density of soil	17.5	17.7	17.74
10	The saturated density of soil	20.48	20.42	20.17
11	Cohesion of soil	27	21	25
12	Internal angle of friction	26.9	27	26.4

Table 3. Characteristics Properties of soil at Dasgaon- Position 2 through GNSS - Latitude- N18° 6′ 48″ and Longitude- E 73° 21′ 58″

Sr. No	Parameters	Sample A	Sample B	Sample C
1	Natural moisture content	21	24	23
2	Specific gravity	2.21	2.24	2.17
3	% of soil passing through 200 No. Sieve	36	34	33.8
4	Liquid limit	51.8	54.47	52.03
5	Plastic limit	41.78	43.23	43.67
6	Plasticity index	10.02	11.24	8.36
7	Optimum moisture content	1.58	1.51	1.46
8	Maximum dry density	27.3	24.1	20.76
9	The dry density of soil	18.390	18.301	18.792
10	The saturated density of soil	20.86	20.91	20
11	Cohesion of soil	30	26	29
12	Internal angle of friction	24	21	22

3.3   Use of GNSS to Study the Same Object for Position Change Observations

The survey has been carried out using GNSS receivers to study the changes if any for the predefined object in the vicinity of ERS, and it has been noted that the position of object was displaced against the initial one noted on a first field visit. The initial position was (Position-1) Latitude- N18° 6′ 46.08″ and Longitude- E 73° 21′ 54″; later it is (Position-2) Latitude- N18° 6′ 48″ and Longitude- E 73° 21′ 58″. These minute changes were noted and the soil sample was collected for finding out input design parameters at real-time study. The changes in soil characteristics properties were observed and its impact on serviceability was studied in detail.

4   Different Geotechnical Properties of Soils and Their Impacts on Input Design Parameters of ERS

4.1   Effect of the Percentage of Finer Material on the Shear Strength Parameter of Soil

Due to weathering action, the percentage of finer material in the composition of the soil is going to increase. The percentage of finer material in the composition of soil may affect on shear strength parameter of soil which affects the calculation factor of safety while design. The Graphs A, B and C, D shows the effect of percentage of finer material on the internal angle of friction and cohesion of backfill material collected from Dasgaon for two different positions observed respectively. The direct shear test is used to find out this relationship.

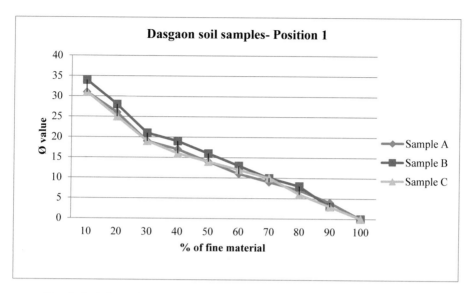

Graph A. Effect of percentage of finer material on the internal angle of friction

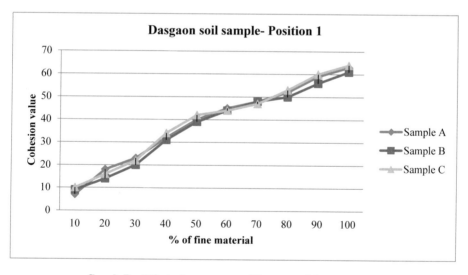

Graph B. Effect of percentage of finer material on cohesion

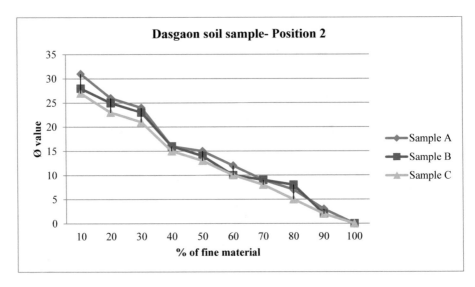

Graph C. Effect of percentage of finer material on the internal angle of friction

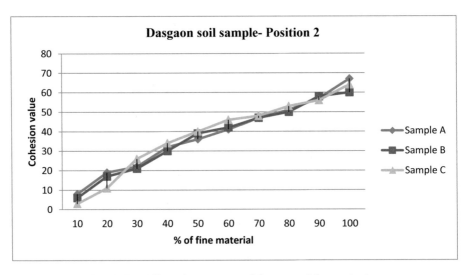

Graph D. Effect of percentage of finer material on cohesion

From this graph, it is concluded that as a percentage of finer material in soil composition goes on increasing the internal angle of friction is going to decrease and cohesion value is going to increase that means in terms of finer material composition, cohesion and internal angle of friction are inversely proportional.

4.2   Effect of the Percentage of Natural Moisture Content on the Shear Strength Parameter of Soil

Majorly retaining structure was failed in the rainy season; therefore it triggers to need to compute changes in shear strength parameters at the full saturated condition. The natural moisture content in soil effects on shear strength parameter and shear strength and which may affect the factor of safety. The Graphs E, F, and G, H shows the effect of Natural moisture content on the internal angle of friction and cohesion of backfill material collected from Dasgaon for two different positions respectively. The percentage of finer material is kept as it is described in Tables 2 and 3. The direct shear test is used to find out the changing effect of shear strength parameter.

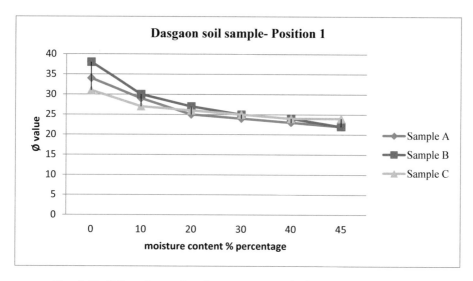

Graph E. Effect of natural moisture content on the internal angle of friction

From this graph, it is concluded that a percentage of natural moisture content in soil composition goes on increasing then both cohesion and internal angle of friction of soil is going to decrease.

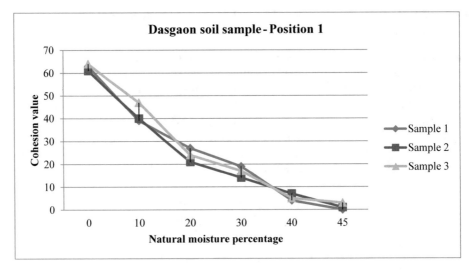

Graph F. Effect of percentage of natural moisture content on cohesion

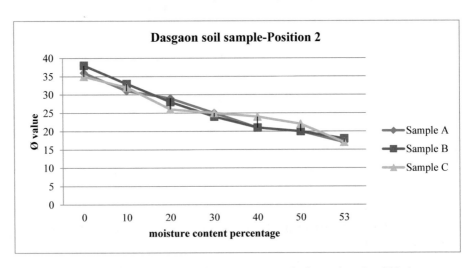

Graph G. Effect of natural moisture content on the internal angle of friction

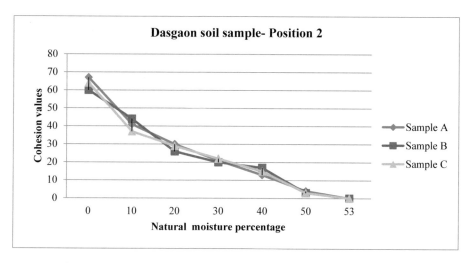

Graph H. Effect of percentage of natural moisture content on cohesion

4.3   Saturated Density at the Full Liquid Limit Condition of Soil

The saturated density is higher than that of dry density of soil and is the combination of soil mass and water. The saturated density is higher at the liquid limit of that soil and is occurred in the rainy season. Therefore the saturated density is calculated for every collected soil sample at their respective liquid limit. The Table 4 shows the maximum saturated density at their respective liquid limit for both positions.

Table 4. Values of liquid limit at their liquid limit

Sr. No.	Area of soil sample collected	Saturated density (Kn/M^2)		
		Sample 1	Sample 2	Sample 3
1	Dasgaon- Position 1	21.45	21.80	20.90
2	Dasgaon- Position 2	22.21	22.47	22.44

5   Study of Serviceability of ERS

5.1   Effect of Current Input Design Parameter on Design Serviceability in Terms of Factor of Safety

At fully saturated condition the cohesion is becoming negligible and minimum internal angle of friction becomes 22° and 17° for Dasgaon - Position 1 and Dasgaon – Position 2 backfill material respectively. The maximum saturated density also at the full liquid limit becomes 21.80, 22.47 KN/M^3 for Dasgaon - Position 1 and Dasgaon – Position 2 backfill material respectively. Height of retaining structure, Soil bearing capacity and properties of concrete and steel is kept as it is. Then the revised design of structure and changes in FOS are noted. The Table 5 shows the Factor of safety calculated at the fully saturated condition.

Table 5. Factor of safeties with different design condition

Sr. No.	Factor of safety	Dasgaon-Position 1			Dasgaon-Position 2		
		At the time of Pre-design	At current Natural moisture condition	At fully saturated condition	At the time of Pre-design	At current Natural moisture condition	At fully saturated condition
1	Overturning	3.43	2.71	2.41	3.84	2.71	2.18
2	Sliding	2.54	1.96	1.40	2.09	1.96	1.128
3	Minimum pressure at the base	74.24	53.30	34.06	56.57	53.30	–93.06
4	Maximum pressure at the base	143.48	174.95	209.28	128.49	174.95	248.76

6   Conclusion

The performance and reliability of Earth Retaining Structures (ERS) deteriorate with respect to time due to changes in characteristics geotechnical properties of soil due to severe environmental conditions such as heavy rainfall, humidity, maximum and minimum temperature, changes in physical and chemical properties of backfill material, changes in Land Use and Land Cover, resulting in effective influence in service life by displacing of ERS from its original position with potential loss of the economy.

Thus, it becomes necessary to consider the serviceability of ERS with these varying design parameters during the effective service period for proper planning and management. This can be achieved using a continuous monitoring system for precise positioning study, which will help in getting early warning related to displacement of ERS from its original position.

In this paper, the effect of changing input design parameters on serviceability of earth retaining structures was observed by studying cantilever retaining structures under different conditions with position change study using GNSS Technology.

The influence of several parameters such as natural moisture contents, the specific gravity of soil, Atterberg's limits, maximum dry density and optimum moisture content, cohesion and internal angle of friction are examined. A comparative study is carried out which shows the time-dependent changes in input design parameters which affect the design specifications, that impacts the serviceability of ERS.

Thus, the comparative study shows the need to monitor and control the input design parameters considered in the design of ERS with some early warning system, continuously to enjoy design service life of ERS effectively using Modern tool of GNSS Technology.

References

1. Turner, A.K.: Social and environmental impacts of landslides. Innovative Infrast. Solutions J. (2018). Springer https://doi.org/10.1007/s41062-018-0175-y
2. Gandomi, A.H., et al.: Optimization of retaining wall design using recent swarm intelligence techniques. Eng. Struct. (2015). https://doi.org/10.1016/j.engstruct.2015.08.034
3. Goh, A.T.C.: Behavior of cantilever retaining walls. J. Geotechn. Eng. (1993). https://doi.org/10.1061/(ASCE)0733-9410(1993)119:11(1751)
4. Udomchai, A., et al.: Failure of riverbank protection structure and remedial approach: A case study in Suraburi province, Thailand. Eng. Fail. Anal. (2018). https://doi.org/10.1016/j.engfailanal.2018.04.040
5. Ukritchon, B., et al.: Optimal design of reinforced concrete cantilever retaining walls considering the requirement of slope stability. KSCE J. Civ. Eng. (2017). https://doi.org/10.1007/s12205-017-1627-1
6. Ellirtgwood, B.: Design and construction error effects on structural reliability. J. struct. Eng. (1987). https://doi.org/10.1061/(asce)0733-9445(1987)113:2(409)
7. Butler, C.J., et al.: Retaining wall field condition, inspection, rating analysis, and condition assessment. J. Perform. Constructed Facil. (2016). https://doi.org/10.1061/(asce)cf.1943-5509.0000785
8. Sharma, C., et al.: Evaluation of the effect of lateral soil pressure on cantilever retaining wall with soil type variation. IOSR J. Mech. Civ. Eng. (2014). https://doi.org/10.9790/1684-11233642
9. Castillo, E., et al.: Design and sensitivity analysis using probability safety factor, an application to retaining wall. Struct. Saf. (2004). https://doi.org/10.1016/s0167-4730(03)00039-0
10. Sivakumar Babu, G.L., Munwar Basha, B.: Optimum design of cantilever retaining walls using target reliability approach. Int. J. Geomech. 8(4), 240–252 (2008)
11. Adunoye, G.O.: Fines content and angle of internal friction of a lateritic soil: an experimental study. Am. J. Eng. Res. (2014). http://www.ajer.org/papers/v3(3)/C0331621.pdf
12. Giacheti, H.L., et al.: Seasonal influence on cone penetration test: an unsaturated soil site example. J. Rock Mech. Geotechn. Eng. (2019). https://doi.org/10.1016/j.jrmge.2018.10.005
13. Zevgolis, I.E., et al.: Probabilistic analysis of retaining walls. Comput. Geotech. (2010). https://doi.org/10.1016/j.compgeo.2009.12.003
14. Collin, J.G.: Lessons learns from segmental retaining wall failure. Geotext. Geomembr. (2001). https://doi.org/10.1016/s0266-1144(01)00016-4
15. Small, J.C., et al.: Structural integrity issues associated with soils and rock in civil engineering industries. Module Mater. Sci. Mater. Eng. (2018). https://doi.org/10.1016/b0-08-043749-4/01147-2
16. Blahova, K., et al.: Influence of water content on the shear strength parameters of clayey soil in relation to stability analysis of a hillside in BRNO region (2013). https://doi.org/10.11118/actaun201361061583
17. Manzari, M.T., et al.: Significance of soil dilatancy in slope stability analysis. J. Geotech. Geoenviron. Eng. (2000). https://doi.org/10.1061/(asce)1090-0241(2000)126:1(75)
18. Abdullahi, M.M.: Evaluation of causes of retaining wall failure (2009). http://lejpt.academicdirect.org/A14/011_018.pdf
19. Ghosh, R.: Effect of soil moisture in the analysis of undrained shear strength of compacted clayey soil. J. Civ. Eng. Constr. Technol. (2013). http://worldcat.org/issn/21412634
20. Xu, S.-Y., et al.: Analysis of the stress distribution across a retaining wall backfill. Comput. Geotech. (2018). https://doi.org/10.1016/j.compgeo.2018.07.001

21. Bobade, S.S., et al.: Study and analysis of causative factors of slumping for designing the preventive measures: a case study in South Konkan, India. Int. J. Comput. Appl. (2012). https://www.ijcaonline.org/proceedings/icett/number2/9838-1019
22. Roy, S., et al.: Role of geotechnical properties of soil on civil engineering structures. Res. Environ. (2017). https://doi.org/10.5923/j.re.20170704.03
23. Zou, Y., et al.: Angle of internal friction and cohesion of consolidated ground marigold petals. Am. Soc. Agric. Eng. (2001). https://doi.org/10.13031/2013.6419
24. Mor, Y., et al.: Reliability-based service-life assessment of aging concrete structures. J. Struct. Eng. (1993). https://doi.org/10.1061/(asce)0733-9445(1993)119:5(1600)
25. Jimoh, Y.A.: Shear strength/moisture content models for a laterite soil in Ilorin, Kwara State, Nigeria (2006). https://doi.org/10.3233/978-1-61499-656-9-521
26. Wu, Y., et al.: Effect of soil variability on bearing capacity accounting for non-stationary characteristics of undrained shear strength. Comput. Geotech. (2019). https://doi.org/10.1016/j.compgeo.2019.02.003
27. Xu, Y., et al.: Determination of peak and ultimate shear strength parameters of compacted clay. Eng. Geol. (2018). https://doi.org/10.1016/j.enggeo.2018.07.001
28. Wei, Y., et al.: The effect of water content on the shear strength characteristics of granitic soils in South China. Soil Tillage Res. (2018). https://doi.org/10.1016/j.still.2018.11.013

Feasibility Study of Bagasse Ash as a Filling Material

A. K. Bhoi[(⊠)], J. N. Mandal, and A. Juneja

Civil Engineering Department, Indian Institute of Technology Bombay,
Powai, Mumbai, Maharashtra 400076, India
{bhoi.aditya, ajuneja}@iitb.ac.in,
cejnm@civil.iitb.ac.in

Abstract. India is the second most sugarcane producer in the world and generates 10 million tons of bagasse ash every year. Bagasse ash is generally spread as fertilizer in the field. It is the most frequent method of disposing of bagasse ash. However, it contains heavy metals which may lead to adverse effect on the yielding of the crop. Hence, some scholars recommend not using bagasse ash as fertilizer. Previous studies indicated that bagasse ash has been significantly used as a fine aggregate in concrete. As a fine aggregate bagasse ash also has the potential to be an alternative filling material. However, a comprehensive characterization of bagasse ash as an alternative filling material is significantly lacking. The present study aims at the characterization of bagasse ash as an alternative filling material instead of natural material. A series of direct shear test and permeability test were conducted for this purpose. The effect of water content and dry density on the shear strength parameter and permeability were studied. The test results show that angle of internal friction and apparent cohesion increase up to optimum moisture content and decreased after wards. The permeability of test specimen decreased with an increase in dry density. As a result, bagasse ash is comparable with conventional fill material.

1   Introduction

Sugarcane production in India is about 300 million tons per year (Purohit and Michaelowa 2007). Bagasse used as a fuel in same industries, which generate about 10 million tons of bagasse ash annually (Ganesan et al. 2007; Srinivasan and Sathiya 2010). If we can utilize this huge amount of by-product, then we will save a huge amount of natural resources, as well as save our environment from being polluted. Spreadings bagasse ash in the fields as a fertilizer has been the usual means of disposal (Lee et al. 1965). Lima et al. (2009) indicated the occurrence of heavy metals in bagasse ash is more than permissible limit. As the bagasse ash contain heavy metal and its application as fertilizer may leads to accumulation of heavy metal, which may leads to some adverse condition, i.e. yielding of crop may lower down (Saxena and Misra 2010). Hence some scholar recommend not to use bagssse ash as fertilizer (Saxena and Misra 2010; Lima et al. 2009). Some scholars in the past studies the use of bagasse ash as a substitute pozzolanic material in concrete (Amin 2011; Amin and Alam 2011; Bahurudeen et al. 2014; Bahurudeen et al. 2015; Chusilp et al. 2009a; Cordeiro et al. 2009; Fairbairn et al.

© Springer Nature Switzerland AG 2020
L. Hoyos and H. Shehata (Eds.): GeoMEast 2019, SUCI, pp. 81–94, 2020.
https://doi.org/10.1007/978-3-030-34206-7_7

2010; Frías et al. 2011; Ganesan et al. 2007; Rajasekar et al. 2018; Rukzon and Chindaprasirt 2012; Shafiq et al. 2016; Singh et al. 2000; Srinivasan and Sathiya 2010; Sua-lam and Makul 2013; Villar-Cocina et al. 2013). In this concern many scholars have studied the effectiveness of bagasse ash as a stabilizing agent for expansive soil (Barasa et al. 2015; Gandhi 2012; Kumar Yadav et al. 2017; Moses and Osinubi 2013; Murari et al. 2015; Sabat 2012). Recently, some researchers have contributed major research works to baggase ash as fine aggregate in concrete (Aigbodion et al. 2010; Almeida et al. 2015; Arif et al. 2016; Bilir et al. 2015; Modani and Vyawahare 2013; Moretti et al. 2016; Prusty et al. 2016; Purohit and Michaelowa 2007; Rashad 2016; Sales and Lima 2010; Shafigh et al. 2014); raising the prospect that assessment has been made to use it as a filling material. Many attempts have been made by several scholars to utilize fly ash as a filling material to reclaim lands (Kim and Chun 1994), and as a backfill material in retaining walls and slope (Nadaf and Mandal 2016; Ram Rathan Lal and Mandal 2012; Ram Rathan Lal and Mandal 2014; Ram Rathan Lal et al. 2015). IRC:SP: 58 (2001) and IRC:SP: 102 (2014) suggest use of fly ash whenever possible. Bagasse ash (BA) can be used for above purposes as it possesses similar properties.

The literature reviewed shows that there is considerable amount of studies have been done on fly ash as a fill material and IRC:SP: 58 (2001) and IRC:SP: 102 (2014) also recommend the use of fly ash as a substitute of natural fill material. Some studies also show the use of bagasse ash as pozzolanic material, fine aggregate in concrete and soil stabilizing agent. But the literature reveals that there are no studies carried out to investigate the suitability of bagasse ash as a fill material. Since the proposed application of bagasse ash is as filling materials, importance have been given to determine grain size distribution, Atterberg's limits, shear strength, permeability and compaction characteristics.

2   Testing Materials

The bagasse ash was collected in dry form from the Karmaveer Shankarrao Kale Sahakari Sakhar Karkhana Ltd., which is situated at Gautamnagar, Kolpewadi, Kopargaon taluk, Ahmednagar district, Maharashtra, India. Bagasse ash was obtained from the dump yard. The sample was dry.

3   Testing Methods

The material collected from site was dry, but had some galvanized iron wire and lumps. The wire and lumps were removed before characterisation of bagasse ash. When a material used as fill material it need to adhere certain specifications specified in some internationally acclaimed standard (Berg et al. 2009; BS 8002 1994; IRC:SP: 102 2014; IRC:SP: 58 2001). Berg et al. (2009) and IRC: SP: (2014) had specified that evaluation grain size distribution and plasticity is most important, followed by permeability, compaction behaviour of potential fill materials and shear strength parameters.

3.1 Specific Gravity

Five representative samples were tested for specific gravity. The specific gravity test followed the ASTM D854 (2014) (method A) standard.

3.2 Grain Size Analysis

Sieve analysis and hydrometer tests were performed in the laboratory to determine the grain size distribution of the bagasse ash. Particle sizes above 75 micron and below 75 micron were separated by wet sieve analysis before testing. The grain size analysis tests followed the ASTM D7928 (2017a) and ASTM D6913 (2017b) standard.

3.3 Microscopic Examination

The surface characteristic of the bagasse ash was examined using a Scanning Electron Microscope (SEM) at Sophisticated Analytical Instrument Facility (SAIF) I.I.T. Bombay.

3.4 Atterberg's Limits

Atterberg's limits were performed in the laboratory to determine the plastic limit (PL) and liquid limit (LL) of the bagasse ash. These Atterberg's limits tests followed the ASTM D 4318 (2017c) standard. Cone penetration method was employed for determination of liquid limit (IS: 2720 (Part 5) 1985).

3.5 Compaction

IRC:SP: 102 (2014) suggest compaction properties should be determined from heavy compaction. Hence, premeasured bagasse ash and water were mixed uniformly by hand in a big tray, followed by compaction tests. The modified proctor compaction tests followed the ASTM D1557 (2012) standard.

3.6 Free Swelling Index

Free swelling index of soils test was performed in the laboratory to determine the increase in volume of the bagasse ash, without any peripheral restrictions when submerged in water. The free swelling index of soils test followed the IS 2720 (Part XL) (1977) standard.

3.7 Consolidation

Consolidation properties test was performed on sample prepared at optimum moisture content and 95% of maximum dry density in the laboratory to determine the compression index (Cc) and recompression index (Cr) of the bagasse ash. The consolidation properties test followed the ASTM D2435 (2011) standard.

3.8   Permeability

Permeability tests were performed in the laboratory to determine the coefficient of permeability of the bagasse ash. The tests were planned to examine effect of water content and dry density on the permeability. The falling head permeability test followed the ASTM D5856 (2015) standard.

3.9   Direct Shear Test

Direct shear tests were performed in the laboratory to determine the shear strength parameter of the bagasse ash. The direct shear test followed the IS 2720 (13) (1986) standard. The tests were planned to examine the effect of dry density and moisture content of the bagasse ash on the shear strength parameters, under modified compactive effort. The tests were performed with dry density and moisture contents along the compaction curve ranging from dry side of optimum to wet side of the optimum.

3.10   pH

The pH tests were performed in the laboratory to determine the acidic and alkaline characteristics of bagasse ash. The pH test followed the ASTM D4972 (2013) standard. Method B was employed to determine the pH value of bagasse ash.

4   Results and Discussion

4.1   Specific Gravity

The specific gravity of five samples of bagasse ash found out to be 2.28, 2.24, 2.31, 2.34 and 2.1, and 2.25 in average. The variation in the specific gravity may be due to variation of vesicular textures of bagasse ash. Bagasse ash with solid structure would have higher apparent specific gravity compare to bagasse ash with more pores. The ranges of specific gravity of the bagasse ash samples were comparable to specific gravity of bagasse ash available in literature. Table 1 shows a comparative statement of specific gravity of bagasse ash proposed for this study, specific gravity of bagasse ash available in literature and specific gravity of fly ash recommended by IRC:SP: 58 (2001) standard. Specific gravity of bagasse ash found out to be very similar to that of fine aggregate.

Table 1. Specific gravity of bagasse ash and fine aggregate

Authors	Material	Specific gravity
Fill proposed for present study	Bagasse ash	2.1–2.34
IRC:SP: 58 (2001)	Fly ash/Pond Ash	1.90–2.55
Sales and Lima (2010)	Bagasse ash	2.23–2.65

(*continued*)

Table 1. (*continued*)

Authors	Material	Specific gravity
Sales and Lima (2010)	Natural sand	2.11
Prusty et al. (2016)	Bagasse ash	1.25–2.54
Prusty et al. (2016)	Sand	2.38–2.64
Almeida et al. (2015)	Bagasse ash	2.57
Almeida et al. (2015)	Sand	2.45
Moretti et al. (2016)	Bagasse ash	2.60
Moretti et al. (2016)	Quartz sand	2.33
Arif et al. (2016)	Bagasse ash	1.95
Modani and Vyawahare (2013)	Bagasse ash	1.25
Modani and Vyawahare (2013)	River sand	2.64
Chusilp et al. (2009a)	Bagasse ash	2.08–2.50
Chusilp et al. (2009b)	Bagasse ash	2.20
Cordeiro et al. (2012)	Bagasse ash	2.53
Bahurudeen et al. (2014)	Bagasse ash	1.91–2.12
Ganesan et al. (2007)	Bagasse ash	1.85
Bahurudeen and Santhanam (2015)	Bagasse ash	1.91
Bahurudeen et al. (2015)	Bagasse ash	2.1
Sua-iam and Makul (2013)	Bagasse ash	2.35
Amin (2011)	Bagasse ash	1.80

4.2   Grain Size Analysis

Figure 1 shows the grain size distribution curve of two sample of bagasse ash tested in laboratory along with grain size distribution of bagasse ash available in literature. The sample 1 had 0.00% gravel, 45.47% sand, 48.06% silt and 6.47% clay size particle. The sample 2 had 0.00% gravel, 42.00% sand, 52.37% silt and 5.63% clay size particle. Table 2 shows the comparative statement between particle size distributions of proposed bagasse ash and particle size distributions recommend by IRC:SP: 58 (2001). The material was found to be uniformly graded with coefficient of uniformity (Cu) equal to 1.8 for sample 1 and 5 for sample 2. Similarly coefficient of curvature (Cc) equal to 0.94 for sample 1 and 1.51 for sample 2. According to Indian standard the soil was classified as sandy silt (SM-ML). Sales and Lima (2010) classified the Brazilian sugarcane bagasse ash as fine sand. Many literature (Bahurudeen et al. 2014; Cordeiro et al. 2009; Sua-Iam and Makul 2013) indicate that bagasse ash comprises nearly equal percentage fine sand and silt size particle. The ranges of particle size of the proposed bagasse ash were little bit higher than the ranges of particle size of bagasse ash available in literature, but comparable to the bagasse ash tested by Bahurudeen et al. (2014), Sua-Iam and Makul (2013) and Cordeiro et al. (2009).

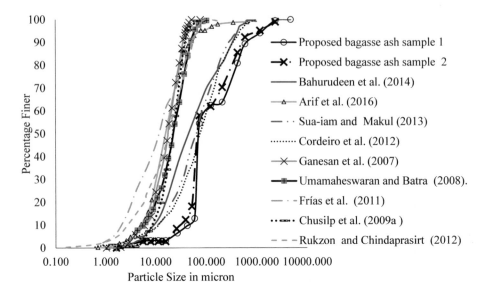

Fig. 1. Grain size distribution of bagasse ash

Table 2. Particle size distribution compare with IRC:SP: 58 (2001)

Particle size	Materials fraction of proposed bagasse ash (%)		Materials fraction of fly ash recommended by IRC:SP:58 (2001) (%)
	Sample 1	Sample 2	
Clay size	6.47	5.63	1–10
Silt size	48.06	52.37	8–85
Sand size	45.47	42.00	7–90
Gravel size	0.00	0.00	0–10
Coefficient of uniformity	1.8	5	3.1–10.7

4.3   Microscopic Examination

The image obtained at 100000 x magnification is shown in Fig. 2. The bagasse ash had large fibrous particle similar to irregular tubular-shaped with large amount of porous. The surface of the bagasse ash had high porosity with rough surface.

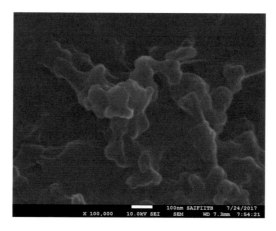

Fig. 2. SEM microphotographs of bagasse ash

4.4 Atterberg's Limits

Cone penetration method was employed for determination of liquid limit, but the bagasse ash started bleeding during the test. As a result of which, its liquid limit could not be measured. Similarly 3.2 mm thread could not be made by hand rolling for plastic limit. Hence the bagasse ash was classified as non plastic soil. It is satisfying Plasticity Index criteria (Plasticity Index should be equal to or below 6) suggested by Berg et al. (2009) and IRC:SP: 102 (2014).

4.5 Compaction

Figure 3.a shows the compaction curve of bagasse ash curve under modified compactive effort. Here it was observed that dry density 1.06 g/cm^3 was achieved with 0% moisture content and dry density went up to 1.15 g/cm^3 with 24% of water content. The further addition of water content after 24% moisture content result in bleeding and subsequently dry density started decreasing. After optimum moisture content bagasse ash was unable to hold the water, as a result of which bleeding started (see Fig. 3.b). The optimum moisture content and maximum dry density of bagasse ash found out to be 24% and 1.15 g/cm^3 respectively. IRC:SP: 58 (2001) recommend maximum dry density between 0.9 to 1.6 g/cm^3 and optimum moisture content between 38 to 18% for fly ash.

(a)   **(b)**

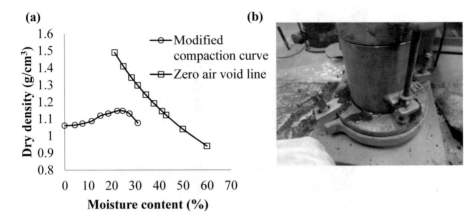

Fig. 3. **a.** Compaction curve. **b.** Bleeding of bagasse ash after optimum moisture content

4.6   Free Swelling Index

The free swelling index of bagasse ash found out to be 0.

4.7   Consolidation

Bagasse ash sample for consolidation test was made with 25% moisture content and 1.09 g/cm³ dry density (95% of maximum dry density). Figure 4 shows e-log(σ_v) curve of bagasse ash. The e-log v curve was used to compute compressibility parameters compression indexes (Cc) and recompression index (Cr). The average value of Cc is 0.092 and 0.18 when the effective stress is in the range of 10 to 200 kPa and 200 to 800 kPa respectively. The unloading part of the curve was used for calculation of average value of Cr and the Cr was found out to be 0.016.

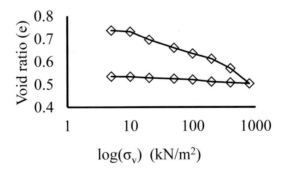

Fig. 4.   e-log(σ_v) curve of bagasse ash

Ram Rathan Lal and Mandal (2012) reported the Cc value of Koradi fly ash equal to 0.078 and 0.092 when the applied effective stress is less than 100 kPa and more than 100 kPa respectively, and average Cr equal to 0.014.

Kaniraj and Gayathri (2004) reported the Cc value of Dadri fly ash equal to 0.041 when effective stress less than 300 kPa and 0.084 when effective stress is greater than 300 kPa. Proposed bagasse ash's Cc and Cr value are nearly similar to fly ash.

4.8   Permeability

Bagasse ash sample for consolidation tests were prepared with following moisture contents and dry density: 4.39% and 1.06 g/cm³, 7.63% and 1.07 g/cm³, 11.56% and 1.09 g/cm³, 15.3% and 1.12 g/cm³, 18.64% and 1.13 g/cm³, 22.49% and 1.15 g/cm³, 24.87% and 1.15 g/cm³, 27.72% and 1.13 g/cm³, and 31.19% and 1.07 g/cm³. Figure 5 shows the influence of moisture content on coefficient of permeability of bagasse ash. From the Fig. 5, it was observed that the coefficient of permeability of bagasse ash compacted at optimum moisture content with maximum dry density was lower than any other combination of moisture contents and dry density. And also the coefficient of permeability at wet side of optimum moisture content was little bit lower than dry side of optimum moisture (coefficient of permeability was 2.07×10^{-5} cm/s and 2.08×10^{-5} cm/s when compacted with dry density of 1.07 g/cm³ with 31.19% water content and dry density of 1.07 g/cm³ with 7.63% water content respectively) Which fall in the range of 8×10^{-6} to 7×10^{-4} cm/s, as recommended by IRC:SP: 58 (2001).

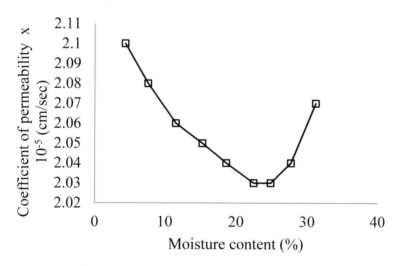

Fig. 5. Variation of coefficient of permeability values with moisture content

4.9   Direct Shear Test

The tests were planned to examine the effect of dry density and moisture content of the bagasse ash on the shear strength parameters, under modified compactive effort.

The tests were performed with dry density and moisture contents along the compaction curve ranging from dry side of optimum to wet side of the optimum. Test specimens were prepared with moisture content and dry density as mentioned in Table 3. Table 3 shows the influence of moisture content and dry density on shear strength parameters of soil. From the Table 3, it was observed that soil compacted at dry side of optimum moisture content possesses the higher angle of internal friction (i.e. 35.68°, 40.55°, 38.55°) compare to soil compacted at wet side of moisture content(i.e. 33.6°, 34.48°). Similarly, bagasse ash compacted dry side of optimum moisture content possess lower apparent cohesion (i.e. 1.65 kN/m^2 and 3.31 kN/m^2), compare to bagasse ash compacted at wet side of moisture (4.95 kN/m^2). IRC:SP: 58 (2001) recommended the range of angle of internal friction from 30° to 40° and negligible cohesion.

Table 3. Influence of moisture content and dry density on shear strength parameters

Sample no.	Dry density (g/cm^3)	Moisture content (%)	Angle of internal friction (φ), (°)	Apparent cohesion (C), (kN/m^2)
1.	1.12	15.3	35.68	1.65
2.	1.13	18.64	40.55	3.31
3.	1.15	22.49	38.55	3.31
4.	1.13	27.72	33.6	4.95
5.	1.07	31.19	34.48	4.95

4.10   pH

The pH value of the bagasse ash was found to be 8. IRC:SP: 58 (2001) recommended the range of pH from 5 to 10. Similarly, Berg et al. (2009) recommended the range of pH from 5 to 10 for reinforced fills with steel reinforcement, 3 to 9 with polyester (PET) reinforcement and pH more than 3 with Polyolefin (PP and HDPE) reinforcements.

5   Conclusion

In this feasibility study, laboratory tests were conducted on bagasse ash consisting of irregular tubular-shaped particle, whose grain were dominantly consist of silt and fine sand size particle. The bagasse ash was non plastic in nature. The SEM microphotographs of bagasse ash show large fibrous particle similar to irregular tubular-shaped with a large amount of porous. This high porosity was facilitating coefficient of permeability of 2.3 × 10^{-5} cm/s. The coefficient of permeability of fly ash should fall in the range of 8 × 10^{-6} to 7 × 10^{-4} cm/s, as recommended by IRC:SP: 58 (2001). The results may be used for selection of appropriate water content and compactive effort for compaction, by considering consolidation parameter, drainage parameter and strength parameters. The modified compaction test result shows dry density 1.06 g/cm^3 was achieved with a moisture content of 0% and a further increase in water content up to 24% gave rise to the maximum dry density 1.15 g/cm^3, followed by a decrease in dry density. On the wet side of compaction curve bleeding is common. That's mean bagasse ash has a low affinity towards the water on the wet side of compaction curve. The angle of internal friction is the main governing factor of shear strength. The dry side of optimum has a higher angle of internal friction compare to wet side of optimum. The compression index (Cc) of bagasse ash is 0.092 when the effective stress is in the range of 10 to 200 kPa and 0.18 when effective stress is in the range of 200 to 800 kPa. The recompression index (Cr) of bagasse ash is 0.016. The compression index (Cc) of bagasse ash was low; hence the deformations due to loadings will be less. The pH value of the bagasse ash was 8; this satisfying pH value criterion for reinforced fills. The results obtained from this study, suggest that bagasse ash satisfy the criterion for fill material laid down by Berg et al. (2009) and IRC:SP: 102 (2014) and can be used as fill material, if suitable design and construction procedure followed.

References

Aigbodion, V.S., Hassan, S.B., Ause, T., Nyior, G.B.: Potential utilization of solid waste (Bagasse Ash). J. Miner. Mater. Charact. Eng. **9**(1), 67–77 (2010)

Almeida, F.C.R., Sales, A., Moretti, J.P., Mendes, P.C.D.: Sugarcane bagasse ash sand (SBAS): brazilian agroindustrial by-product for use in mortar. Constr. Build. Mater. **82**, 31–38 (2015)

Amin, N.: Use of bagasse ash in cement and its impact on the strength and chloride resistivity. J. Mater. Civ. Eng. **23**(5), 717–720 (2011)

Amin, N., Alam, S.: Activation of bagasse ash in cement using different techniques. Proc. Inst. Civ. Eng. Constr. Mater. **164**(4), 199–204 (2011)

Arif, E., Clark, M.W., Lake, N.: Sugar cane bagasse ash from a high efficiency co-generation boiler: applications in cement and mortar production. Constr. Build. Mater. **128**, 287–297 (2016)

ASTM: Standard test methods for one-dimensional consolidation properties of soils using incremental loading. ASTM **D2435**, 1–15 (2011)

ASTM: Standard test methods for laboratory compaction characteristics of soil using modified effort (56,000ft-lbf/ft3 (2,700 kN-m/m3)). ASTM **D1557**, 1–14 (2012)

ASTM: Standard test method for pH of soils. ASTM **D4972**, 1–4 (2013)

ASTM: D854 - standard test methods for specific gravity of soil solids by water pycnometer. ASTM **D854**, 1–8 (2014)

ASTM: Standard test method for measurement of hydraulic conductivity of porous material using a rigid-wall, compaction-mold permeameter. ASTM **D5856**, 1–9 (2015)

ASTM: Standard test method for particle-size distribution (gradation) of fine-grained soils using the sedimentation (hydrometer) analysis. ASTM **D7928**, 1–25 (2017a)

ASTM: Standard test methods for particle-size distribution (gradation) of soils using sieve analysis. ASTM **D6913**, 1–34 (2017b)

ASTM: Standard test methods for liquid limit, plastic limit, and plasticity index of soils. ASTM **D4318**, 1–20 (2017c)

Bahurudeen, A., Marckson, A.V., Kishore, A., Santhanam, M.: Development of sugarcane bagasse ash based Portland pozzolana cement and evaluation of compatibility with superplasticizers. Constr. Build. Mater. **68**, 465–475 (2014)

Bahurudeen, A., Santhanam, M.: Influence of different processing methods on the pozzolanic performance of sugarcane bagasse ash. Cement Concr. Compos. **56**, 32–45 (2015)

Bahurudeen, A., Kanraj, D., Gokul Dev, V., Santhanam, M.: Performance evaluation of sugarcane bagasse ash blended cement in concrete. Cement Concr. Compos. **59**, 77–88 (2015)

Barasa, P.K., Jonah, K., Mulei, S.M.: Stabilization of expansive clay using lime and sugarcane bagasse ash. Int. J. Sci. Res. (IJSR) **4**(4), 2112–2117 (2015)

Berg, R., Christopher, B., Samtani, N.: Design and Construction of Mechanically Stabilized Earth Walls and Reinforced Soil Slopes–Volume I. Federal High Way Administration (FHWA), Washington, D.C. (2009)

Bilir, T., Gencel, O., Topcu, I.B.: Properties of mortars with fly ash as fine aggregate. Constr. Build. Mater. **93**, 782–789 (2015)

BS 8002: Code of Practice for Earth Retaining Structures. British Standards Institution, London (1994)

Chusilp, N., Jaturapitakkul, C., Kiattikomol, K.: Utilization of bagasse ash as a pozzolanic material in concrete. Constr. Build. Mater. **23**(11), 3352–3358 (2009a)

Chusilp, N., Jaturapitakkul, C., Kiattikomol, K.: Effects of LOI of ground bagasse ash on the compressive strength and sulfate resistance of mortars. Constr. Build. Mater. **23**(12), 3523–3531 (2009b)

Cordeiro, G.C., Toledo Filho, R.D., Tavares, L.M., Fairbairn, E.de M.R.: Ultrafine grinding of sugar cane bagasse ash for application as pozzolanic admixture in concrete. Cem. Concr. Res. **39**(2), 110–115 (2009)

Fairbairn, E.M.R., Americano, B.B., Cordeiro, G.C., Paula, T.P., Toledo Filho, R.D., Silvoso, M. M.: Cement replacement by sugar cane bagasse ash: CO2 emissions reduction and potential for carbon credits. J. Environ. Manag. **91**(9), 1864–1871 (2010)

Frías, M., Villar, E., Savastano, H.: Brazilian sugar cane bagasse ashes from the cogeneration industry as active pozzolans for cement manufacture. Cement Concr. Compos. **33**(4), 490–496 (2011)

Gandhi, K.S.: Expansive soil stabilization using bagasse ash. Int. Jo. Eng. Res. Technol. **1**(5), 5–7 (2012)

Ganesan, K., Rajagopal, K., Thangavel, K.: Evaluation of bagasse ash as supplementary cementitious material. Cement Concr. Compos. **29**(6), 515–524 (2007)

IRC:SP: 58: Guidelines for Use of Fly Ash in Road Embankments. Indian Roads Congress, New Delhi (2001)

IRC:SP: 102: Guidelines for Design and Construction of Reinforced Soil Walls. Indian Roads Congress, New Delhi (2014)

IS: 2720 (Part 5): Determination of Liquid and Plastic Limit. Indian Standards Institute, New Delhi (1985)

IS 2720 (Part XL): Determination of Free Swell Index of Soils. Bureau of Indian Standards, New Delhi (1977)

IS 2720 (13): Direct Shear Test. Bureau of Indian Standards, New Delhi (1986)

Kaniraj, S.R., Gayathri, V.: Permeability and consolidation characteristics of compacted fly ash. J. Energy Eng. **130**(1), 18–43 (2004)

Kim, S.S., Chun, B.S. (1994): The study on a practical use of wasted coal fly ash for coastal reclamation. In: 13th International Conference on Soil Mechanics and Foundation Engineering, pp. 1607–1612. CRC Press, London (1994)

Krishna, A., Latha, G.: Modeling the dynamic response of wrap-faced reinforced soil retaining walls. Int. J. Geomech. **12**(August), 439–450 (2011)

Kumar Yadav, A., Gaurav, K., Kishor, R., Suman, S.K.: Stabilization of alluvial soil for subgrade using rice husk ash, sugarcane bagasse ash and cow dung ash for rural roads. Int. J. Pavement Res. Technol. **10**(3), 254–261 (2017)

Lee, L.H., et al. (1965): The application of bagasse furnace ash to sugar cane fields, Rept. Taiwan Sugar Expt. Sta. (Taiwan), pp. 38, 53–79 (1965)

Lima, S.A., Sales, A., Santos, T.J. (2009): Caracterização físico-química da cinza do bagaço da cana-de-açúcar visando o seu uso em argamassas e concretos como substituto do agregado miúdo (Physicochemical characterization of the sugarcane bagasse ash for using in mortars and concretes as a natural aggregate replacement). 51°. Congresso Brasileiro do Concreto (Brazilian Congress Concrete). Proceedings. IBRACON, São Paulo (2009)

Modani, P.O., Vyawahare, M.R.: Utilization of bagasse ash as a partial replacement of fine aggregate in concrete. Procedia Eng. 51(NUiCONE 2012), 25–29 (2013)

Moretti, J.P., Sales, A., Almeida, F.C.R., Rezende, M.A.M., Gromboni, P.P.: Joint use of construction waste (CW) and sugarcane bagasse ash sand (SBAS) in concrete. Constr. Build. Mater. 113, 317–323 (2016)

Moses, G., Osinubi, K.J.: Influence of compactive efforts on cement- bagasse ash treatment of expansive black cotton soil. Int. J. Civ. Environ. Struct. Constr. Architect. Eng. 7(7), 1541–1548 (2013)

Murari, A., Singh, I., Agarwal, N., Kumar, A.: Stabilization of Local Soil with Bagasse Ash (April), pp. 37–39 (2015)

Nadaf, M.B., Mandal, J.N.: Steel grid reinforced fly ash slopes. In: Proceedings of Geo-Chicago 2016, ASCE, Chicago, pp. 678–687 (2016)

Prusty, J.K., Patro, S.K., Basarkar, S.S.: Concrete using agro-waste as fine aggregate for sustainable built environment – a review. Int. J. Sustain. Built Enviro. Gulf Organ. Res. Dev. 5(2), 312–333 (2016)

Purohit, P., Michaelowa, A.: CDM potential of bagasse cogeneration in India. Energy Policy 35 (10), 4779–4798 (2007)

Rajasekar, A., Arunachalam, K., Kottaisamy, M., Saraswathy, V.: Durability characteristics of Ultra High Strength Concrete with treated sugarcane bagasse ash. Constr. Buil. Mater. 171, 350–356 (2018)

Ram Rathan Lal, B., Mandal, J.N.: Feasibility study on fly ash as backfill material in cellular reinforced walls. Electron. J. Geotech. Eng. 17(J), 1637–1658 (2012)

Ram Rathan Lal, B., Mandal, J.N.: Behavior of cellular-reinforced fly-ash walls under strip loading. J. Hazard. Toxic Radioact. Waste 18(1), 45–55 (2014)

Ram Rathan Lal, B., Padade, A.H., Mandal, J.N. (2015): Effect of single and double anchored systems on the behaviour of cellular reinforced fly ash walls. In: IFCEE 2015 © ASCE 2015, GSP 256, pp. 1473–1482 (2015)

Rashad, A.: Cementitious materials and agricultural wastes as natural fine aggregate replacement in conventional mortar and concrete. J. Build. Eng. 5, 119–141 (2016)

Rukzon, S., Chindaprasirt, P.: Utilization of bagasse ash in high-strength concrete. Mater. Des. 34, 45–50 (2012)

Sales, A., Lima, S.A.: Use of Brazilian sugarcane bagasse ash in concrete as sand replacement. Waste Manag 30(6), 1114–1122 (2010)

Saxena, P., Misra, N.: Remediation of heavy metal contaminated tropical land. In: Sherameti, I., Varma, A. (eds.) Soil Heavy Metals, Soil Biology, vol. 19, pp. 431–477. Springer, Berlin (2010)

Sabat, A.K.: Utilization of bagasse ash and lime sludge for construction of flexible pavements in expansive soil areas. Electr. J. Geotech. Eng. 17 H, 1037–1046 (2012)

Shafigh, P., Mahmud, H.Bin, Jumaat, M.Z., Zargar, M.: Agricultural wastes as aggregate in concrete mixtures - A review. Constr. Build. Mater. 53, 110–117 (2014)

Shafiq, N., Hussein, A.A.E., Nuruddin, M.F., Mattarneh, H.Al.: Effects of sugarcane bagasse ash on the properties of concrete. In: Proceedings of the Institution of Civil Engineers – Engineering Sustainability, pp. 1–10 (2016)

Singh, N.B., Singh, V.D., Rai, S.: Hydration of bagasse ash-blended portland cement. Cem. Concr. Res. **30**(9), 1485–1488 (2000)

Srinivasan, R., Sathiya, K.: Experimental study on bagasse ash in concrete. Int. J. Serv. Learn. Eng. **5**(2), 60–66 (2010)

Sua-Iam, G., Makul, N.: Use of increasing amounts of bagasse ash waste to produce self-compacting concrete by adding limestone powder waste. J. Cleaner Prod. **57**, 308–319 (2013)

Villar-Cocina, E., Frias, M., Hernandez-Ruiz, J., Savastano Jr., H.: Pozzolanic behaviour of a bagasse ash from the boiler of a Cuban sugar factory. Adv. Cem. Res. **25**(3), 136–142 (2013)

Development of a Constitutive Model for Clays Based on Disturbed State Concept and Its Application to Simulate Pile Installation and Setup

Firouz Rosti[1] and Murad Abu-Farsakh[2(✉)]

[1] Department of Chemical, Civil and Mechanical Engineering,
McNeese State University, Lake Charles, LA 70605, USA
[2] Louisiana Transportation Research Center, Louisiana State University,
Baton Rouge, LA 70808, USA
cefars@lsu.edu

Abstract. In this paper, an elastoplastic model is proposed to describe the behavior of clayey soils subject to disturbance at the soil-structure interaction for application to pile installation and the following setup. The soil remolding that occurs during deep penetration and the following soils thixotropic strength regaining over time were modeled in this study. The disturbed state concept (DSC) was used as a core of the proposed model, and the critical state theory was adopted to define the main components for the DSC model. The Modified Cam-Clay (MCC) model was implemented within the context of DSC to define the intact state response. A novel approach was applied to define the soil shear response for the MCC model to have it applicable in DSC. Furthermore, the soil remolding during shear loading was related to the deviatoric plastic strain developed in the soil body. The proposed model, referred as Critical State and Disturbed State Concept (CSDSC) model, can capture the elastoplastic behavior of both NC and OC soils. The proposed model was implemented in Abaqus software, and it was then validated using the triaxial test results available in the literatures. Very good agreement was obtained between the triaxial test results and the CSDSC model prediction for different stress paths, stress-strain response and the generated excess porewater pressures. Furthermore, pile installation and the following pile setup behavior were modeled using the proposed CSDSC model. The predicted values for pile resistance using CSDSC model were compared with the values measured from field load tests, which indicated that the proposed model is capable of simulating pile installation and predict appropriately the pile capacity as well as the disturbance behavior at the soil body.

Keywords: Elastoplastic constitutive model · Soil thixotropy · CSDSC · Disturbed state concept · Critical state theory · Hardening and softening · Triaxial test

© Springer Nature Switzerland AG 2020
L. Hoyos and H. Shehata (Eds.): GeoMEast 2019, SUCI, pp. 95–118, 2020.
https://doi.org/10.1007/978-3-030-34206-7_8

1 Introduction

Soils experience variety of load histories during natural deposition, which results in soils to be over-consolidated (OC) or normally consolidated (NC). In engineering application, OC and NC soils experience different behavior under applied external loads; the OC soils exhibit more complicated behavior and have lower void ratio and higher shear strength (Yao et al. 2007). In cases, such as deep penetration, the soil disturbance and particle remolding occur during shear loading and significantly affect the general soil behavior. For engineering problems that involve deep foundations and piles, the soil type usually changes with depth due to difference in the stress history. Therefore, in such a case, incorporating an appropriate constitutive model that can capture the actual behavior for both NC and OC soils is necessary. There are several elastoplastic constitutive model available in the literature attempted to model the soil response under different loading conditions. Most of the developed constitutive models for clays are based on the critical state soil mechanics (CSSM) concept (Pestana and Whittle 1999). The critical state concept models had been formulated based on laboratory test results in axisymmetric condition. The modified Cam-Clay (MCC) model proposed by Roscoe and Burland (1968) is the most well-known critical state model. The MCC model is able to appropriately describe the isotropic NC clay behavior. Since the MCC model indicates elastic response inside the yield surface, its prediction for OC clay is poor (Likitlersuang 2003). Researchers developed series of bounding surface models to overcome this deficiency (e.g., Dafalias and Herman 1986). The bounding surface plasticity concept was used later to develop the MIT-E3 model by Whittle (1993). The bounding surface plasticity has been developed to provide smooth transition from elastic to fully plastic state for soils under general loading. In the bounding surface model, the hardening parameter depends on the distance from the current stress state to an imaginary stress at the bounding surface. The application of the critical sate models for heavily OC clays is limited, and it needs specific consideration. Yao et al. (2007) and (2012) introduced a unified hardening model using Hvorslev envelope to capture the heavily OC clay behavior. Linear and parabolic form of the Hvorslev envelope were adopted to adjust the conventional MCC model for heavily OC clay response under shear loads. Based on CSSM and bounding surface theory, Chakraborty et al. (2013a, b) developed a two surface elastoplastic constitutive model to capture strain rate dependent behavior for clay. Chakraborty et al. (2013a, b) and Basu et al. (2014) used two-surface plasticity constitutive model for the clays, and it was implemented for analysis of shaft resistance in piles. Although, their models were able to describe both NC and OC clay behavior, but they usually include plenty of material parameters that requires performing several lab tests. Furthermore, these models are not able to describe the disturbance occurs in soil body under shear loading in the elastoplastic formulation. Likitlersuang (2003) introduced a rate dependent version for the hyperelasticity model, in which he verified the proposed model by simulating triaxial test results performed on the Bangkok clay. Zhang et al. (2014) introduced a mathematical model to explain the soil shear behavior at the pile interface by using a hyperbolic and bi-linear relation between the skin friction and the relative displacement between pile surface and adjacent soil. They used a different hyperbolic relation to define the softening behavior between the unit skin friction and the pile-soil relative displacement.

Soil disturbance is obvious in the soil body subjected to external shear loading, which causes shear failure in the soil, such as the case in deep penetration problems. Therefore, an appropriate constitutive model should be able to consider the actual soil behavior due to disturbance occurred in the soil-structure interface during shear loading for application to deep penetration problems, such as pile installation. The disturbed state concept (DSC) developed by Desai and Ma (1992) is a powerful technique that is directly formulated based the soil disturbance. In the DSC model, the soil response is obtained using two boundary (reference) state responses, which are the relative intact (RI) state and the fully adjusted (FA) or critical (c) state. The real or observed soil behavior is defined as average combination of RI and FA responses. An appropriate elastoplastic constitutive model usually used to describe the soil behavior of the RI state; while the critical state is used to describe the FA state response of soils. On other words, it is assumed that the soil is in critical state condition when it reaches to FA state. Desai and his coworkers used the elastoplastic hierarchical single surface (HISS) model to define the intact state response (e.g., Desai et al. 1986; Wathugala 1990; Shao 1998; Pal and Wathugala 1999; Katti and Desai 1999; Desai 2005 and Desai 2007; Desai et al. 2011). Hu and Pu (2003) used the conventional hyperbolic model to define the intact reference state, in which they developed an elastoplastic constitutive model for sandy soil response at the soil-structure interface.

For geotechnical cases that deal with extreme shear loading, remolding of soil particles and strength regaining over time after end of soil disturbance due to the soil thixotropic response is common. Fakharian et al. (2013) described a reduction factor to incorporate soil remolding in the numerical simulation. Barnes (1997) introduced a time-dependent exponential function to formulate inks thixotropic response after remolding. In this paper, a new constitutive model is developed based on the combination of the critical state theory and the DSC, which can describe the actual soil behavior of both NC and OC clay soils and the soil disturbance caused by shear loading during deep penetration. In this study, the thixotropic response of clay soils was formulated similar to the inks thixotropy presented in Barnes (1997). The proposed model requires only six model parameters, which is less than the models that have been previously developed based on the DSC. Furthermore, it predicts smooth transition from elastic to plastic state of soil under shear loading, which is usually observed during laboratory tests performed on soil samples.

2 Modified Cam-Clay (MCC) Model

The original Cam-Clay model was developed based on the CSSM concept by Roscoe and Schofield (1963), which was then modified by Roscoe and Burland (1968) to develop the MCC model. This model was developed to study the clayey soil behavior under applied loads. The MCC model has been used widely in past decades to define the soil behavior because, firstly, it is capable in capturing soil realistic behavior than the other conventional models such as Mohr-Coulomb or Von-Misses models.

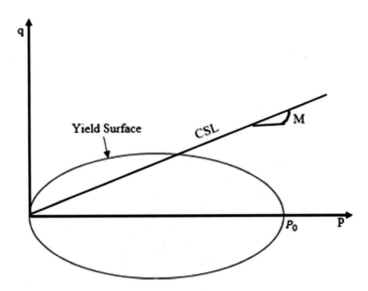

Fig. 1. Elliptical yield surface for MCC model in $p' - q$ plane.

Secondly, it is a simple model with pretty less parameters in comparison with other advanced soil models. The MCC yield locus is assumed to have an elliptical shape, as shown in Fig. 1, and the equation of the yield surface in triaxial stress space is given as:

$$F = q^2 - M^2 \left[p' (p'_0 - p') \right] = 0 \tag{1}$$

where p' is the general volumetric stress or hydrostatic stress, which is $p' = (\sigma'_1 + 2\sigma'_3)/3$ for triaxial stress state. q is the deviatoric stress, which is $q = \sigma'_1 - \sigma'_3$ for triaxial condition. M is the slope of critical state line on p-q plane, and p'_0 is the pre-consolidation pressure. The Cam Clay model was developed based on the volumetric behavior of the saturated soil under shear loading, unloading and reloading. For MCC model, the following differential equations define the material behavior:

$$d\sigma = C^{ep} d\varepsilon \tag{2}$$

where $d\sigma$ is the incremental stress tensor, $d\varepsilon$ is the incremental strain tensor, C^{ep} is the elastoplastic constitutive matrix, and it can be obtained from Eq. (3).

$$C^{ep}_{ijkl} = C^e_{ijkl} - \frac{C^e_{ijtu} \frac{\partial F}{\partial \sigma_{tu}} \frac{\partial F}{\partial \sigma_{pq}} C^e_{pqkl}}{H + \frac{\partial F}{\partial \sigma_{mn}} C^e_{mnrs} \frac{\partial F}{\partial \sigma_{rs}}} \tag{3}$$

where C^e is the elastic stress-strain matrix and depends on the stress state through the elastic bulk modulus K and the shear modulus G. The term H is the hardening parameter that can be obtained using the following equations:

$$H = \frac{1+e}{\lambda - \kappa} p' p'_0 \frac{\partial F}{\partial p'}$$

(4)

where e is the soil void ratio; and λ and κ represent the slope of the normally consolidated line and the unloading-reloading line, respectively.

3 Disturbed State Concept (DSC)

The disturbed state concept is defined based on the disturbance occurs in the soil body during external loading especially shear loads. The conceptual framework of disturbed state concept is based on the cyclical behavior of materials from its "cosmic state" to the engineering materials state, and then the tendency to return to its initial cosmic state under applied loads (Desai 2001). When the soil deforms under applied loads, its initially intact fabric undergoes microstructural changes, such as sliding, reorientation and cracking of the soil particles (Katti and Desai 1995). By proceeding deformation, part of the soil body become disturbed, which then changes to the fully adjusted (FA) or critical (c) state, while the rest part is still in relative intact (RI) state condition. These two reference states (FA and RI) are used to define observed (averaged) soil behavior under applied loads. Based on this concept, the observed or averaged (a) behavior of the material is an average combination of the RI and FA based on the amount of disturbance occurred in the soil body. An appropriate constitutive model such as linear or nonlinear elastic and elastoplastic model can define the RI response (Desai 2001). The real behavior of a material is a combination of two interacting behaviors in the RI and FA reference states. The stresses at the average state (σ^a) can be obtained from a linear combination of the RI and FA states and by using the disturbance function D according to the following relation:

$$\sigma^a_{ij} = (1 - D)\sigma^i_{ij} + D\sigma^c_{ij}$$

(5)

where the notations a, i, and c are representing the observed, intact and fully adjusted (critical state) responses, respectively. At the initial stage of applying load, the RI response has more influence on the overall behavior of a material, and with continuing the applied load, the material particles are displacing and the overall material response gradually approaches to the FA state. The disturbance function D, which is related to the produced plastic strain in the soil body under shear is used to define the averaged response. The following exponential equation was proposed by Desai (2001), which relates soil disturbance D to the developed plastic strain in the soil body under applied load:

$$D = 1 - e^{-A*\xi_d^B} \tag{6}$$

where ξ_d is the trajectory of deviatoric plastic strain dE_{ij}^p, defined as $\xi_d = \int \left(dE_{ij}^p . dE_{ij}^p \right)^{1/2}$; and A and B are material parameters, which are obtained from results of laboratory soil triaxial tests. At the beginning of loading ($\xi_d = 0$), the function D represents the initial disturbance under natural condition, which is usually assumed to be equal to zero (i.e., soil is undisturbed at initial stage). Then with proceeding at shear loading, plastic strain increases, and hence the soil transform gradually from intact state to fully disturbed state, and therefore, the D function approaches to unity.

For a specific yield surface F and in the case of associated flow rule used for plasticity formulation, the plastic strain increment is related to the deferential of the yield function F with respect to the stress tensor by the following equation:

$$d\varepsilon_{ij}^p = \lambda^* \frac{\partial F}{\partial \sigma_{ij}} \tag{7}$$

where λ^* is the plastic multiplier. The star sign in λ^* is used here for plastic multiplier to remove confusion. The incremental deviatoric strain dE_{ij}^p can be obtained as:

$$dE_{ij}^p = d\varepsilon_{ij}^p - \frac{1}{3} d\varepsilon_{kk}^p \delta_{ij} = \lambda^* \left(\frac{\partial F}{\partial \sigma_{ij}} - \frac{1}{3} \frac{\partial F}{\partial \sigma_{kk}} \delta_{ij} \right) \tag{8}$$

The plastic strain increment trajectory $d\xi_d$ will be obtained as:

$$d\xi_d = \left(dE_{ij}^p . dE_{ij}^p \right)^{1/2} = \lambda^* \left[\frac{\partial F}{\partial \sigma_{ij}} \frac{\partial F}{\partial \sigma_{ij}} - \frac{1}{3} \frac{\partial F}{\partial \sigma_{kk}} \frac{\partial F}{\partial \sigma_{ll}} \right]^{1/2} \tag{9}$$

By combining Eqs. (6) and (9), the incremental change in the disturbance function is obtained as follows:

$$dD = AB\xi_d^{B-1} e^{-A*\xi_d^B} . \lambda^* . \left[\frac{\partial F}{\partial \sigma_{ij}} \frac{\partial F}{\partial \sigma_{ij}} - \frac{1}{3} \frac{\partial F}{\partial \sigma_{kk}} \frac{\partial F}{\partial \sigma_{ll}} \right]^{1/2} \tag{10}$$

In Eq. (10), λ^* is obtained from applying equilibrium and consistency conditions for a specific yield surface.

4 Proposed Constitutive Model

In this study, the intact state behavior was modeled using the MCC model. For fully adjusted response, again, the critical state soil mechanics concept was adopted, and it was assumed that the soil is in critical state when it becomes fully adjusted. Adopting the critical state concept for both RI and FA behaviors imposes the model to have two different critical state parameters (M) in order to describe the behaviors of RI and FA reference states: the value for intact material response (M_i) and the value for the fully adjusted response (M_c). The former one is not a real soil property, and its value defined based on the proposed model requirement (as will be described in the following sections). The latter one indicates the critical state parameter of soil, and its value is obtained from laboratory test results. Since the proposed model was obtained by combining the DSC and the critical state MCC model, it is called the Critical State and Disturbed State Concept (CSDSC) model. In this model, the disturbance function D is applied to the critical state parameter M as follows:

$$M_a = (1 - D)M_i + DM_c \tag{11}$$

where M_a is the averaged or observed value for the critical state parameter at each stage of loading process. Figure 2 describes that the evolution of the critical state parameter M_a during shear loading. At the initial stage of shear, the soil is assumed to be undisturbed ($D = 0$ and $\xi_d = 0$), which means Eq. 11 yields to $M_a = M_i$. However, with the proceeding of the applied load, the soil disturbs, the plastic strains develop in the soil body, the values of D and ξ_d increase, and eventually the D value approaches to 1. At this point, the soil reaches the critical state (i.e. $M_a = M_c$) condition.

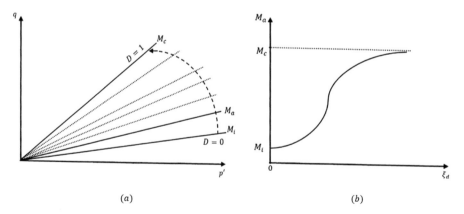

(a)   (b)

Fig. 2. Evolution of critical state parameter M during shear loading

M_c and M_i were assumed to be constant, so the incremental form for Eq. (11) can be expressed as follows:

$$dM_a = (M_c - M_i)dD \tag{12}$$

In case of using the MCC model to represent the intact material response, the plastic multiplier λ^* can be defined as follows:

$$\lambda^* = \frac{\frac{\partial F}{\partial \sigma_{ij}} C^e_{ijkl} d\varepsilon^i_{kl}}{\frac{\partial F}{\partial \sigma_{mn}} C^e_{mnpq} \frac{\partial F}{\partial \sigma_{pq}} - \left[\frac{1+e}{\lambda - \kappa} p' p'_0 \frac{\partial F}{\partial p'}\right]} \tag{13}$$

where $d\varepsilon^i_{kl}$ is the incremental intact strain. Replacing Eq. (13) into Eq. (10) yields to the following equation:

$$dD = \frac{\left(AB\xi^{B-1}_d e^{-A*\xi^B_d}\right) \frac{\partial F}{\partial \sigma_{ij}} . C^e_{ijkl} \left[\frac{\partial F}{\partial \sigma_{rs}} \frac{\partial F}{\partial \sigma_{rs}} - \frac{1}{3}\frac{\partial F}{\partial \sigma_{uu}} \frac{\partial F}{\partial \sigma_{vv}}\right]^{1/2}}{\frac{\partial F}{\partial \sigma_{mn}} C^e_{mnpq} \frac{\partial F}{\partial \sigma_{pq}} - \left[\frac{1+e}{\lambda - \kappa} p' p'_0 \frac{\partial F}{\partial p'}\right]} . d\varepsilon^i_{kl} \tag{14}$$

Equation (14) indicates that the incremental change in the disturbance function in related strain increment of the intact material. By combining Eqs. (12) and (14), the incremental change for observed critical state parameter dM_a, is obtained as follows:

$$dM_a = (M_c - M_i)\frac{\left(AB\xi^{B-1}_d e^{-A*\xi^B_d}\right) \frac{\partial F}{\partial \sigma_{ij}} . C^e_{ijkl} \left[\frac{\partial F}{\partial \sigma_{rs}} \frac{\partial F}{\partial \sigma_{rs}} - \frac{1}{3}\frac{\partial F}{\partial \sigma_{uu}} \frac{\partial F}{\partial \sigma_{vv}}\right]^{1/2}}{\frac{\partial F}{\partial \sigma_{mn}} C^e_{mnpq} \frac{\partial F}{\partial \sigma_{pq}} - \left[\frac{1+e}{\lambda - \kappa} p' p_0 \frac{\partial F}{\partial p'}\right]} . d\varepsilon^i_{kl} \tag{15}$$

The elastic matrix C^e is related to the nonlinear elastic bulk and shear modulus K and G. As indicated by Zhang [11] and Sloan [26], the tangent modulus K and G cannot be used directly in numerical analysis of critical state models because of nonlinearity of them within the finite strain increment. Therefore, the secant modulus can be used to replace the tangent modulus as follows:

$$\bar{K} = \frac{p'}{\Delta \varepsilon^e_v} \left(\exp\left(\frac{1+e}{\kappa} \Delta \varepsilon^e_v\right) - 1\right) \tag{16}$$

where p' is the effective hydrostatic stress at the start of the volumetric elastic strain increment $\Delta \varepsilon^e_v$. By assuming that the Poisson' ratio stays constant during loading; the secant shear modulus can be obtained as:

$$\bar{G} = \frac{3(1 - 2v)}{2(1 + v)} \bar{K} \tag{17}$$

In the proposed model, MCC model runs in each increment (or sub-increment). However, while the soil shears, the critical state parameter M evolves gradually from

M_i value and to M_c value based on the amount of developed plastic strain in each increment and obeying the DSC theory. Figure 3 presents the formulation of the proposed model in the $p' - q$ space. The point A represents the stress state at the beginning of the strain increment $d\varepsilon_n$. The MCC model is used to solve the governing equations for $d\varepsilon_n$ using M_a^n, and the new stress state is obtained at point B, which is located on the yield surface a_n. Then, updated value for the averaged critical state parameter M_a^{n+1} is obtained from the incremental value of dM_a by using Eq. (15) for use in the next increment. The imaginary yield surface i_{n+1} will then be defined using the updated critical state parameter M_a^{n+1} and the hardening parameter p_c^{n+1} (the prime index in p_c' removed for simplicity). The current stress state (point B) is located inside the imaginary yield surface i_{n+1}, which causes the elastoplastic behavior for the material in the next steps until stress state reaches the critical state. The MCC model is then solved using the new strain increment $d\varepsilon_{n+1}$ to reach point C and so on. The main advantage of this approach is the possibility of specifying a small value close to zero for M_i since the observed behavior is captured by the disturbance parameters regardless of the chosen value for M_i. By choosing a very small value for M_i, the plastic behavior inside the yield surface is achieved; leading to a smooth transition between the elastic and plastic behavior. For each strain increment,$d\varepsilon$, the elastic and the plastic portions are determined using the yield surface intersection parameter α_{inter} as follows:

$$d\varepsilon^e = \alpha_{inter}.d\varepsilon \tag{18a}$$

$$d\varepsilon^p = (1 - \alpha_{inter}).d\varepsilon \tag{18b}$$

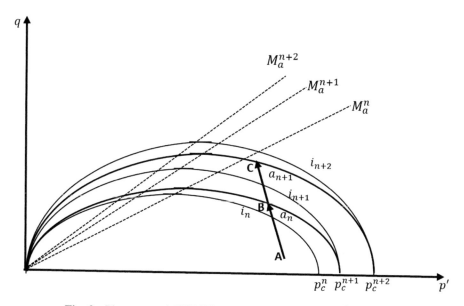

Fig. 3. The proposed (CSDSC) model representation in $p' - q$ space

In Eq. (18a, 18b), higher values of α_{inter} indicates dominant elastic response, while lower α_{inter} shows dominant plastic response. A value of $\alpha_{inter} = 0$ indicates that under strain increment $d\varepsilon$ pure plastic deformation occurs, while a value of $\alpha_{inter} = 1$ represents pure elastic deformation. In the proposed model, at the initial stage after loading ($D = 0$), the elastic behavior is dominant (point B is far from the corresponding yield surface a). By proceeding with loading, the α_{inter} value decreases, the elastic portion reduces, and the plastic behavior become dominant until it reaches the fully plastic response at $D = 1$ (point B locates on the yield surface). Summary of the required steps to implement the CSDSC model are described as follows:

1. For a given strain increment $d\varepsilon$, solve the constitutive equations using the MCC model and implement an appropriate integration scheme to determine the current stress state (Point B in Fig. 3) and corresponding p_c.
2. Calculating the disturbance function increment, dD, based on the induced plastic strain values using the described formulation. Then, calculate dM_a using Eq. (15) to update the M_a value for use in the next increment.
3. An imaginary yield surface is defined based on the updated M_a and p_c values. This step causes the current stress states (point B in Fig. 3) to stay inside the imaginary yield surface).
4. Run the MCC model using the imaginary yield surface and the M_a value obtained from step 2, which yields new stress state at point C and new hardening parameter p_c.
5. Repeat Steps 2 to 4 until the stress state reaches the critical state ($M_a = M_c$) condition.

5   CSDSC Model Parameters

The proposed CSDSC model has six parameters, including the following four critical state (MCC) model parameters: (1) The Poisson ratio ν, (2) the slope of the critical state line M, (3) slope of the normal compression line λ, and (4) slope of unloading-reloading line κ; and two disturbed state parameters for defining the disturbance function D. The first four parameters can be obtained directly from laboratory tests such as consolidation and triaxial tests. Several researchers discussed how to obtain these parameters from conventional laboratory tests (e.g., Yao et al. 2007 and Guan-lin and Bin 2016). There are two parameters in the disturbance function D, namely, A and B parameters, which can be obtained from triaxial test results by some mathematical manipulation. As Desai (2001) has indicated, the D can be obtained from a triaxial test result with the following equation:

$$D = \frac{q_i - q_a}{q_i - q_c} \tag{19}$$

where q_i, q_c and q_a are the deviatoric stress for RI, FA and averaged state, respectively. On the other side, by rearranging and taking natural logarithms of the disturbance function, Eq. (6) yields to:

$$\ln(A) + B\ln(\xi_d) = \ln(-\ln(1 - D)) \tag{20}$$

By plotting the value for D obtained from Eq. (19) and using the results of CU triaxial test versus the obtained values for ξ_d, a best fit straight line as shown in Fig. 4 can be used to determine the A and B parameters.

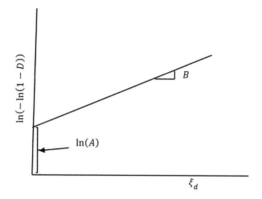

Fig. 4. Determination of disturbance function parameters.

6   Soil Thixotropy and the CSDSC Model

Fine-grained cohesive soils show certain degrees of thixotropic response under constant effective stress and constant void ratio. Thixotropy is defined as the "process of softening caused by remolding, followed by a time-dependent return to the original harder state" (Mitchel 1960). Thixotropy is a reversible process, which is mainly related to the rearrangement of the remolded soil particles, and it must be considered in constitutive models that deal with shear failure at the soil-structure interface such as driven piles, and the following increase in pile capacity with time after end of driving (or pile setup). Therefore, for numerical study of pile installation and setup phenomenon, the thixotropic behavior of cohesive soil should be considered in the model. For driven piles, which deal with change in the soil properties during different steps of installation and the following setup, it is necessary to adopt appropriate material properties at each step. Numerical simulation of pile setup using properties obtained from laboratory tests like triaxial or consolidation tests on undisturbed soil samples yields unrealistic results. Therefore, the time-dependent reduction parameter $\beta(t)$ was applied in this study on the critical state parameter M and the soil-structure interface friction coefficient μ to incorporate the effect of soil remolding during pile installation and the following strength regaining with time after that:

$$\begin{cases} M(t) = \beta(t)M \\ \mu(t) = \beta(t)\mu \end{cases} \tag{21}$$

Based on research performed by Barnes (1997) on the thixotropic strength regaining over time for inks, Abu-Farsakh et al. (2015) proposed the following definition for $\beta(t)$:

$$\beta(t) = \beta(\infty) - [\beta(\infty) - \beta(0)]e^{-\left(\frac{t}{\tau}\right)} \tag{22}$$

Where, the parameter t is time after soil remolding. $\beta(0)$ is the initial value for reduction parameter β immediately after soil shearing ($t = 0$), which its value depends on the degree of remolding occurs in the soil during shear. $\beta(\infty)$ is the β value after long time from soil disturbance (t = ∞); and τ is a time constant that controls the rate of evolution of β. Abu-Farsakh et al. (2015) related τ to the soil t_{90}, which is the time for 90% dissipation of the excess pore water pressure at pile surface.

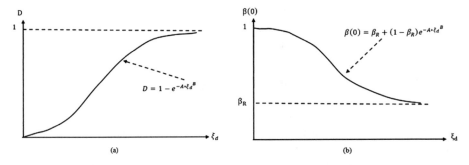

Fig. 5. Variation of soil characteristics during shear loading: (a) disturbance function D, and (b) the soil strength reduction factor immediately after remolding, $\beta(0)$.

The initial value of soil disturbance depends on the amount of disturbance has been developed at soil-pile interface during pile installation. The best parameter which can be used to determine the disturbance is deviatoric strain values due to shear force development at the interface. In this study, a similar formulation to the disturbance function D (i.e. Eq. 6) was also proposed, which relates the initial reduction parameter $\beta(0)$ to the deviatoric plastic strain trajectory using the following exponential function:

$$\beta(0) = \beta_R + (1 - \beta_R)e^{-A*\xi_d^B} \tag{23}$$

where β_R is the β value for the fully remolded soil, which indicates a maximum reduction of the soil strength during shearing, and its value is related to the soil sensitivity. In order to reduce complexity, the disturbed state parameters A and B were used to introduce a relation between $\beta(0)$ and ξ_d in Eq. (23). Figures 5a and b present the schematic representations of the variations of D and $\beta(0)$ versus the deviatoric plastic strain trajectory, respectively. These Figures show that while the soil disturbs, the D value approaches unity, and the $\beta(0)$ yields to β_R by proceeding the plastic strain. Abu-Farsakh et al. (2015) correlated β_R to the soil sensitivity, S_r using a simple relation as $\beta_R = (S_r)^{-0.3}$.

7 Verification of the Proposed Model

To clarify advantages of the proposed CSDSC model, the triaxial compression test for a typical clay was simulated numerically using the Abaqus software. Three-dimensional model with a cubic porous element for soil specimen was used. The coupled porewater pressure analysis was used to define the multi-phase characteristic of the saturated soil. Triaxial stress state was applied using prescribed stresses for confining stress and using the prescribed displacement to apply deviatoric stress. The sample top surface was assumed to be free for drainage. The model was run first using the Abaqus built in MCC model. Then the same model was run using the proposed CSDSC model through implementation via UMAT; and the obtained results were compared with the results of MCC model as shown in Fig. 6. The figure indicates that the MCC model prediction for OC soil is not realistic, which shows mostly elastic response during undrained shearing, especially for heavily OC soils. On the other hand, the proposed CSDSC model provides a rational elastoplastic response with smooth transition from elastic to plastic behavior even for the heavily OC soils.

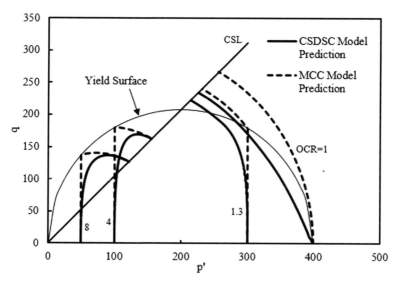

Fig. 6. Comparative result for stress path in triaxial compression obtained from numerical simulation using MCC and the proposed model.

The proposed CSDSC model was then verified via numerical simulation of three case studies. Case1: the triaxial tests performed on Kaolin clay were simulated using CSDSC and MCC models. Case 2: the triaxial test conducted on Boston Blue clay was modeled, and the CSDSC model prediction was compared with the test measurments. Case 3: the proposed CSDSC model was used to simulate a full-scale pile installation and the following pile setup case study of a test pile located at Bayou Laccassine Bridge site, Louisiana. The results then were compared with the measured values obtained from the field load tests. It should be noted that the soil thixotropy formulation (i.e., Eqs. 21 to 23) included only in numerical simulation of the third case because this is the only case that the soil strength regain has been measured through field static and/or dynamic load tests performed after pile installation. However, in the first two case studies, the numerical simulation included only triaxial test procedure using the proposed CSDSC model without considering soil strenght gain after shear because there was no thixotropic information available in the original laboratory tests done by other researchers.

Case study 1: Kaolin Clay

To verify the predictive capability of the proposed model, the results of laboratory triaxial tests on Kaolin clay performed by Yao et al. (2012) was simulated using the proposed CSDSC model. The shear responses from underlined triaxial compression test for different stress history (OCR = 1, 1.20, 5,8,12) were simulated. The four model parameters that are related to the MCC model were obtained from Yao et al. (2012). The remained two model parameters that are related to the disturbed state concept (i.e. A and B) were obtained from triaxial test results and following the procedure outlined by Desai (2001) and explained with Eqs. 19 and 20. After a simple curve fitting to the data obtained from triaxial test results and adopting straight line shown in Fig. 4, values of A = 14.43 and B = 0.47 were obtained. The calculated parameters are presented in Table 1.

Table 1. Model Parameters for Kaolin clay used for implementation (Yao et al. 2012)

M	λ	κ	ν	A	B
1.04	0.14	0.05	0.20	14.43	0.47

Using the model parameters presented in Table 1, the FE model was run with MCC model and the results for different stress paths in the undrained condition are presented in Fig. 7a. The figure shows that the MCC model is not able to capture appropriately the actual soil response under undrained shear loads, especially for OC clays.. In the proposed model, the strong capability of the CSDSC in modeling the actual behavior of soils was demonstrated, and the results of numerical simulation for different stress paths using the proposed model are presented in Fig. 7b. The figure clearly indicates that the proposed model can predict the actual soil behavior for both the NC and OC soils with good agreement. The model is also able to capture the strain softening behavior of heavily OC soils. Figure 8 shows the results of proposed model for stress-strain relation at different over-consolidation ratios, which represents good agreement. In this figure, the stress values are normalized with respect to the initial pre-consolidation pressure p'_0.

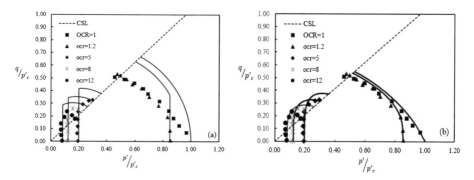

Fig. 7. Prediction of numerical simulation of undrained triaxial test on Kaolin clay (Yao et al. 2012) using (a) MCC model, and (b) the proposed CSDSC model.

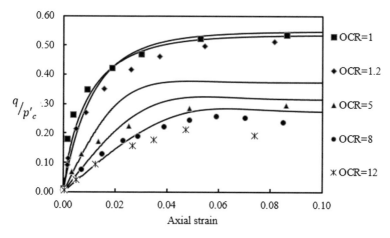

Fig. 8. Stress-strain relation for undrained triaxial test on Kaolin clay (Yao et al. 2012) using the proposed CSDSC model.

Figure 9 presents the numerical simulation for porewater pressure generated during undrained triaxial tests using the proposed CSDSC model, which are in good agreement with the test measurements. The figure shows that, for NC soil and lightly OC soil, the generated porewater pressure is positive, which is representation of soil contraction during undrained shearing. On the other hand, for heavily OC soils, the numerical simulation shows the generation of positive porewater pressure at the initial stage of the test followed by negative pore water pressure until failure. This is an indication of soil dilative behavior, which is common in heavily OC soils. Based on the obtained results, the soil dilation in undrained condition increases by increasing OCR values.

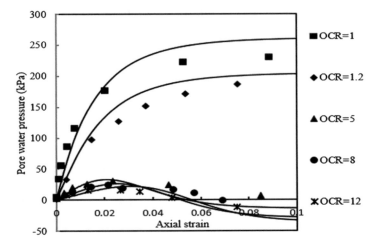

Fig. 9. Prediction of the excess porewater pressure generated in undrained triaxial test on Kaolin clay (Yao et al. 2012) using the proposed CSDSC model.

Case study 2: Boston Blue Clay

The results of undrained triaxial test for normally consolidated Boston Blue Clay, which are available in literature (e.g. Ling et al. 2002), were used to verify the proposed model prediction. The test was simulated using the proposed CSDSC model, and using the soil properties and model parameters as shown in Table 2 (Ling et al. 2002). The required information including stresses, strains and excess porewater pressure were extracted from the numerical model in order to study the variation of these parameters during shearing. Figures 10, 11 and 12 compare the results obtained from the CSDSC model and the corresponding measurements from the triaxial test results. These figures demonstrate very good agreement between the proposed model prediction and the performed triaxial test results, especially in predicting the stress path and the stress-strain relation. However, the CSDSC model prediction for the generated excess pore water pressure during triaxial tests is slightly under estimated, but still within acceptable tolerance.

Table 2. Model Parameters for Boston Blue Clay used for implementation (Ling et al. 2002)

M	λ	κ	ν	A	B
1.39	0.175	0.034	0.23	4.70	0.35

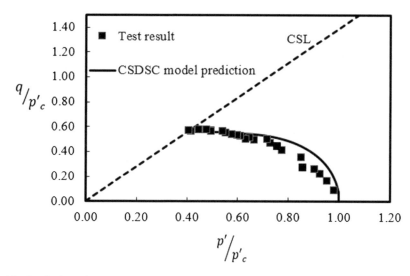

Fig. 10. Prediction of numerical simulation of undrained triaxial test on Boston Blue clay (Ling et al. 2002) using the proposed CSDSC model.

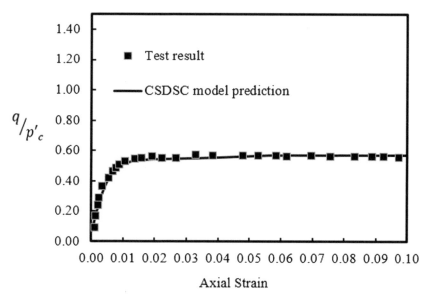

Fig. 11. Stress-strain relation for undrained triaxial test on Boston Blue clay (Ling et al. 2002) using the proposed CSDSC model.

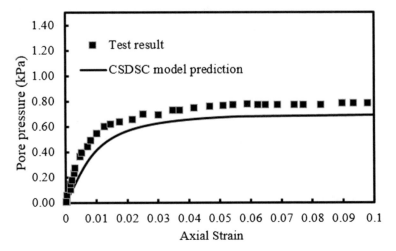

Fig. 12. Prediction of the excess porewater pressure generated in undrained triaxial test on Boston Blue clay (Ling et al. 2002) using the proposed CSDSC model.

Case study 3: Full-Scale Pile Installation

The load test results for a full-scale instrumented test pile that was conducted at Bayou Laccasine Bridge site, Louisiana (Haque et al. 2014) was simulated using the proposed CSDSC model. The test pile was square concrete pile with 0.76 m width and a total length of 22.87 m. A 6.4 m long casing was installed and driven prior to pile installation to represent the scour effect at shallow depth. The test pile was fully instrumented with pressure cells, vibrating wire piezometers and sister bar strain gages that were installed at different depths of pile length, targeting specific soil layers. In addition, the surrounding soils were instrumented with nine multi-level piezometers located at the same depths as the pressure cells and piezometers installed at the pile's face. Both static load tests (SLTs) and dynamic load tests (DLTs) were conducted to obtain the pile resistance at different times after end of driving.

In this paper, the numerical simulation of the pile installation and following setup were performed using the Abaqus software and adopting the techniques described in Abu-Farsakh et al. (2015). The geometry of the soil and pile, the applied boundary conditions, and finite element mesh are shown in Fig. 13. Numerical simulation of pile installation was achieved by applying prescribed displacement to the soil nodes to create volumetric cavity expansion. The pile was then placed inside the cavity followed by applying a vertical penetration until the steady state condition is reached.

The subsurface soil condition at the pile site is mainly consists of clay soil, and the natural water table is 2.24 m below the ground surface. The subsurface soil domain was divided into eight layers based on the soil type and properties as presented in Table 3. In the table, w is the soil water content (%), S_u is the undrained shear strength (kPa), and K is the soil permeability (m/s).

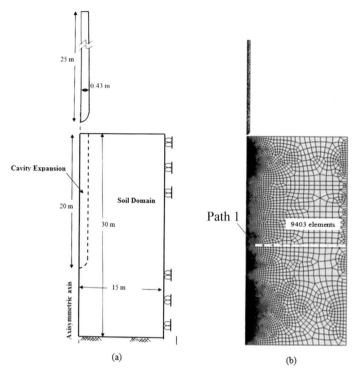

Fig. 13. Numerical simulation domain: (a) geometry and boundary conditions and (b) FE mesh (Abu-Farsakh et al. 2015).

Table 3. Subsurface soil properties for the test pile site (Abu-Farsakh et al. 2015)

Layer No.	Depth (m)	w (%)	S_u (kPa)	OCR	M	λ	κ	K (m/s) 10^{-9}
1	0-6.40	21	120	4	0.61	0.104	0.035	3.80
2	6.40-7.60	26	72	2.5	1.17	0.100	0.029	4.20
3	7.60-10	25	68	2	0.90	0.091	0.026	0.62
4	10-11.60	29	104	1.7	0.90	0.108	0.035	0.12
5	11.60-13	23	94	1.45	0.62	0.108	0.035	7.60
6	13-16	52	150	1.40	1.12	0.147	0.061	8.90
7	16-20	24	112	1.3	0.92	0.100	0.030	0.17
8	20-23	29	101	1	0.93	0.056	0.013	0.66

The proposed CSDSC model was used to describe the elastoplastic behavior of the surrounding clay soil. The soil remolding during pile installation was incorporated in the constitutive model using Eq. (23), and relating β_R to the soil sensitivity S_r with $\beta_R = (S_r)^{-0.3}$. This relation was depicted for Bayou Laccasine Bridge site based on available data for S_r and the pile resistance values obtained from field load tests, which yield a value of $\beta_R = 0.75$ (Abu Fasakh et al. 2015). The disturbed state parameter

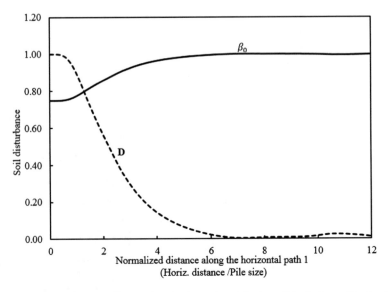

Fig. 14. Variation of β and D for a typical horizontal path in the soil body immediately after pile installation.

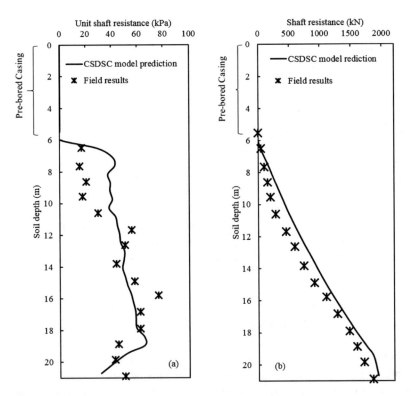

Fig. 15. Comparison between the proposed CSDSC model prediction with measured values from field test results (a) unit shaft resistance, and (b) total shaft resistance.

values of A = 8.50 and B = 0.25 were calculated using laboratory triaxial test results, and were then adopted in the proposed model.

Figure 14 represents the disturbance occurs in the soil immediately after pile installation for a typical horizontal path (path 1 in Fig. 13), which was obtained from numerical simulation using the CSDSC model. The figure shows that β has its maximum value $\beta_R = 0.75$ for soil adjacent to the pile face and approaches unity at a radial distance equal to eight times the pile size. At the same time, the disturbance function has a maximum value ($D = 1$) at the soil-pile interface, and it approaches to $D = 0$ at a radial distance equal to eight times the pile size along the same path.

The numerical simulation using CSDSC model was compared with the measured field test results of the pile load tests. Figure 15a shows the comparison between the predictions of unit shaft resistance one hour after end of driving obtained using the CSDSC model and the measured values from the field load tests. The cumulative values of shaft resistance obtained from numerical simulation were also compared with the calculated values obtained from field tests, and the results are shown in Fig. 15b. These figures clearly indicate that the CSDSC model is able to predict the pile resistance appropriately. For more verification, the increase in pile capacity after end of driving (or pile setup) was obtained from numerical modeling using the CSDSC model, and the model predictions were compared with the measured values from field load test results as shown in Fig. 16. The figure demonstrates that the proposed model is able to simulate pile setup with the model predictions of the pile resistance are slightly over-predicted the measured values.

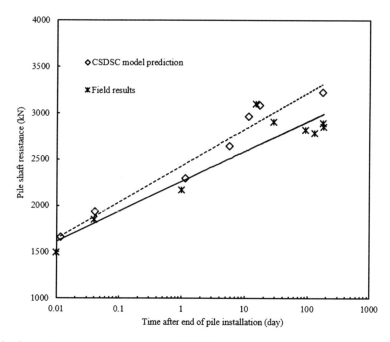

Fig. 16. Comparison between the proposed CSDSC model prediction and field measurement for pile setup.

8 Conclusions

In this paper, an elastoplastic constitutive model for clay soil was developed and evaluated in application of pile installation and the following setup over time. The proposed model is based on the combination of Disturbed State Concept (DSC) and critical state concept of Modified Cam-Clay (MCC) model, which is referred as the CSDSC model. In the CSDSC model, the disturbance function D was applied to the critical state parameter M to adopt disturbed state concept. The soil remolding behavior was related to the state of deviatoric plastic strain developed in the soil body during shear loading for use to simulate deep penetration problems such as pile installation and the corresponding remolding of surrounding soil. The soil thixotropic response was incorporated in the pile setup phenomenon using a time-dependent function, which increases exponentially with time after end of pile driving. The proposed model was implemented in Abaqus software via a user defined subroutine UMAT. The responses of Kaolin and Boston Blue clays under undrained monotonic loads were simulated for verification. The Kaolin clay was evaluated for NC case and OCR values of 1.2, 5, 8 and 12; however, Boston Blue clay was simulated under NC condition. The proposed CSDSC model prediction, which included stress paths, stress-strain curves, and generated excess porewater pressures were compared with the triaxial test results available in the literatures for verification of the model. Furthermore, a full-scale instrumented pile driven in Louisiana clayey soil and the following setup were simulated using the CSDSC model, and the results obtained from FE model and those measured from field test results were compared. Based on the results of this study, the following conclusions can be made for the numerical simulation using the CSDSC model:

(a) The developed CSDSC model has only six parameters, which is less than the previous elastoplastic models developed based on DSC, which makes it more effective in geotechnical engineering applications. The model parameters include four critical state parameters and two parameters related to DSC. All the model parameters can be simply obtained from consolidation and triaxial laboratory tests.

(b) The proposed CSDSC model predictions were compared with laboratory triaxial test results, which show that the model was able to appropriately capture the undrained shear responses for NC clays, lightly OC clays and heavily OC clays.

(c) The comparison between the results of conventional MCC model and the proposed CSDSC model showed that the CSDSC model is able to remove deficiency of the MCC model in simulation the clay soil behavior especially for heavily OC soils. The steep changes in stress paths inside yield surface, which was observed in MCC model, could be vanished in CSDSC model providing a smooth transition from elastic to plastic response. In other word, the CSDSC model is able to capture accurately the smooth transition from elastic to elastoplastic and to fully plastic state, which is usually observed during experimental tests performed on clayey soils.

(d) The numerical simulation of a full-scale pile installation and following pile load tests after end of driving indicated that the proposed CSDSC model is able to simulate pile installation and capture the soil disturbance, soil thixotropy and pile setup appropriately.
(e) The results demonstrated good agreement between the prediction of pile resistance and pile setup using CSDSC model and the measured values obtained from full-scale pile loads tests results.

Acknowledgement. This research is funded by the Louisiana Transportation Research Center (LTRC Project No. 11-2GT) and Louisiana Department of Transportation and Development, LADOTD (State Project No. 736-99-1732).

References

Abu-Farsakh, M., Rosti, F., Souri, A.: Evaluating pile installation and the following thixotropic and consolidation setup by numerical simulation for full scale pile load tests. Can. Geotech. J. **52**, 1–11 (2015)

Barnes, H.A.: Thixotropy-a review. Int. J. Non-Newtonian Fluid Mech. **70**, 1–33 (1997)

Basu, P., Prezzi, M., Salgado, R., Chakraborty, T.: Shaft resistance and setup factors for piles jacked in clay. J. Geotech. Geoenviron. Eng. **140**(3) (2014)

Chakraborty, T., Salgado, R., Loukidis, D.: A two-surface plasticity model for clay. Comput. Geotech. **49**, 170–190 (2013a)

Chakraborty, T., Salgado, R., Basu, P., Prezzi, M.: Shaft resistance of drilled shafts in clay. J. Geotech. Geoinv. Eng. **139**(4), 548–563 (2013b)

Dafalias, Y.F., Herrmann, L.R.: Bounding surface plasticity II: application to isotropic cohesive soils. J. Eng. Mech. **112**(12), 1263–1291 (1986)

Desai, C.S., Somasundaram, S., Frantziskonis, G.: A hierarchical approach for constitutive modeling of geologic materials. Int. J. Num. Anal. Meth. Geomech. **10**, 225–257 (1986)

Desai, C.S., Ma, Y.: Modeling of joints and interface using disturbed state concept. Int. J. Num. Anal. Meth. Geomech. **16**, 623–653 (1992)

Desai, C.S.: Mechanics of Materials and Interface: The Disturbed State Concept. CRC Press, Boca Raton (2001)

Desai, C.S., Sane, S., Jenson, J.: Constitutive modeling including creep and rate-dependent behavior and testing of glacial tills for prediction of motion of glaciers. Int. J. Geomech. **11**, 465–476 (2001)

Desai, C.S.: Constitutive modeling for geologic materials: significance and directions. Int. J. Geomech. **5**, 81–84 (2005)

Desai, C.S.: Unified DSC constitutive model for pavement materials with numerical implementation. Int. J. Geomech. **7**(2), 83–102 (2007)

Desai, C.S., Sane, S., Jenson, J.: Constitutive modeling including creep and rate-dependent behavior and testing of glacial tills for prediction of motion of glaciers. Int. J. Geomech. **11**, 465–476 (2011)

Fakharian, K., Attar, I.H., Haddad, H.: Contributing factors on setup and the effects on pile design parameter. In: Proceedings of 18th International Conference on Soil Mechanics and Geotechnical Engineering, Paris (2013)

Haque, M.N., Abu-Farsakh, M., Chen, Q., Zhang, Z.: A case study on instrumentation and testing full-scale test piles for evaluating set-up phenomenon. In: 93th Transportation Research Board Annual Meeting, vol. 2462, pp. 37–47 (2014)

Hu, L., Pu, J.L.: Application of damage model for soil-structure interface. J. Comput. Geotech. **30**, 165–183 (2003)

Katti, D.R., Desai, C.S.: Modeling and testing of cohesive soils using disturbed-state concept. J. Eng. Mech. **121**, 648–658 (1995)

Likitlersuang, S.: A hyperplasticity model for clay behavior: an application to Bangkok clay. Ph. D dissertation, The University of Oxford (2003)

Ling, H., Yue, D., Kaliakin, V., Themelis, N.: Anisotropic elastoplastic bounding surface model for cohesive soil. J. Eng. Geomech. **7**, 748–758 (2002)

Mitchell, J.K.: Fundamental aspects of thixotropy in soils. J. Soil Mech. Found. Des. **86**, 19–52 (1960)

Pal, S., Wathugala, G.W.: Disturbed state concept for sand-geosynthetic interface and application for pullout test. Int. J. Num. Anal. Meth. Geomech. **23**, 1872–1892 (1999)

Pestana, J.M., Whittle, A.J.: Formulation of a unified constitutive model for clays and sands. Int. J. Num. Anal. Meth. Geomech. **23**, 1215–1243 (1999)

Roscoe, K.H., Schofield, A.N.: Mechanical behavior of an idealized wet clay. In: Proceedings of 2nd European Conference on Soil Mechanics and Foundation Engineering, Wiesbaden, vol. 1, pp. 47–54 (1963)

Roscoe, K.H., Burland, J.B.: On the Generalized Behavior of Wet Clay, Engineering Plasticity, pp. 535–610. Cambridge University Press, Cambridge (1968)

Shao, C.: Implementation of DSC model for dynamic analysis of soil-structure interaction problems. Ph. D. Dissertation. Dept. of Civil Engineering, University of Arizona, Tucson, Arizona (1998)

Sloan, S.W., Abbo, A.J., Sheng, D.: Refined explicit integration of elastoplastic models with automatic error control. Eng. Comput. **18**, 121–154 (2001)

Wathugala, G.W.: Finite element dynamic analysis of nonlinear porous media with application to the piles in saturated clay. Ph. D. Dissertation. Dept. of Civil Engineering, University of Arizona, Tucson, Arizona (1990)

Whittle, A.J.: Evaluation of a constitutive model for overconsolidated clays. Geotechnique **43**(2), 289–313 (1993)

Yao, Y.P., Sun, D.A., Matsuoka, H.: A unified constitutive model for both clay and sand with hardening parameter independent on stress path. J. Comput. Geotech. **35**, 210–222 (2007)

Yao, Y.P., Gao, Z., Zhao, J., Wan, Z.: Modified UH model: constitutive modeling of overconsolidated clays based on a parabolic Hvorslev envelope. J. Geotech. Geoenviron. Eng. **138**, 860–868 (2012)

Guan-lin, Ye, Bin, Ye: Investigation of the overconsolidation and structural behavior of Shanghai clays by element testing and constitutive modeling. Undergr. Space **1**(1), 62–77 (2016)

Zhang, Q., Li, L., Chen, Y.: Analysis of compression pile response using a softening model, a hyperbolic model of skin friction, and a bilinear model of end resistance. J. Eng. Mech. **140**, 102–111 (2014)

Influence of Resting Periods on the Efficiency of Microbially Induced Calcite Precipitation (MICP) in Non-saturated Conditions

Jean-Baptiste Waldschmidt and Benoît Courcelles[(✉)]

Department of Civil, Geological and Mining Engineering,
Polytechnique Montréal, Montreal, Canada
benoit.courcelles@polymtl.ca

Abstract. Microbially Induced Calcite Precipitation (MICP) is a relatively new soil improvement technique that has been extensively studied during the last two decades. Most of these studies have been performed in saturated conditions and only few papers deal with non-saturated conditions. In this study, we investigate two injection protocols with or without resting periods between treatment steps. The results of unconfined compression tests performed on treated samples, as well as the measurement of the calcite contents, lead to the conclusion that the resting periods improve the stiffness of the samples and the calcite content.

1 Introduction

Microbially Induced Calcite Precipitation (MICP), also known as biocalcification, is a relatively new process relying on the formation of precipitated calcium carbonate. This technique is mostly used for granular soils and consists in injecting aerobic bacteria in the soil that can hydrolyze urea to produce carbonate ions. This injection of bacteria is usually followed by an injection of urea, nutrients and a calcium-rich solution. The calcium ions combine with carbonate ions issued from the bacterial activity and lead to the precipitation of calcium carbonate around the bacteria. Ultimately, the precipitation results in the formation of bridges between the grains of soil, such as the natural cementing process of sandstone and other carbonated rocks (Martinez and DeJong 2009; Harkes et al. 2010; Montoya et al. 2012).

A wide variety of parameters must be considered to have a successful reaction, such as the pH, nucleation spots and the availability of calcium and carbonate ions (Stocks-Fischer et al. 1999, Hammes and Verstraete 2002; Girinski 2009). According to the *in-situ* conditions, two different methods can be implemented. The first one, and the most used, consists in injecting directly into the soil all the nutriments in saturated conditions and simultaneously pumping the reactants or biproducts in excess (Montoya et al. 2012; Cheng et al. 2013; Dejong et al. 2013; Martinez et al. 2013; Montoya et al. 2013; Cheng and Cord-Ruwisch 2014). This technique allows below-the-surface and wide spread (until 5 m wide) treatments (van Paassen et al. 2010). The other technique is the injection of all the required reactants by gravity percolation. This technique is not suitable for deep treatment but seems promising for surface level problems as, unlike the previous one, it does not require two sets of wells. Finally, the saturation of the soil is a

© Springer Nature Switzerland AG 2020
L. Hoyos and H. Shehata (Eds.): GeoMEast 2019, SUCI, pp. 119–126, 2020.
https://doi.org/10.1007/978-3-030-34206-7_9

critical factor in the efficiency of the methodology. Using a saturation lower than 100% means that the treatment is more focused on the grains of soil (Cheng et al. 2013).

The objective of this paper is to present two protocols of MICP through gravity percolation and to investigate the influence of resting periods on the efficiency of MICP treatments.

2   Materials and Methods

The general approach of the experiments relies on the treatment of non-saturated soils according to two different protocols, with and without resting periods. The experiments were performed in duplicate to ensure the accuracy of the results.

2.1   Soil Specimens

The treated soil was pure silica sand (Ottawa) with a homogeneous particle-size distribution. This sand is considered poorly graded according to the Unified Soil Classification System (USCS) and its relative density has been evaluated to 2.644 g/cm^3 according to the Canadian National Standard CAN/BNQ 2501-070/2014. The minimum and maximum densities were respectively evaluated to 1.453 and 1.700 g/cm^3 according to the Canadian National Standard CAN/BNQ 2501-062/2005.

2.2   Treatment Solutions

To fulfill the treatment, three kind of solutions must be used: (1) a **bacterial solution**, with a specific concentration of bacteria, (2) a **fixation solution** to facilitate the attachment of bacteria onto the soil's grain and finally, (3) a **cementation solution** to grow the carbonate crystals.

The bacterial strain of Sporosarcina pasteurii (ATCC 11859) was grown on a solid sterile aerobic yeast-extract medium (Tris buffer 15.75 g/L, yeast extract 20 g/L, ammonium sulfate 10 g/L, agar 10 g, pH = 9.0) for 72 h at 30 °C, then put in a fridge at 4 °C for conservation.

For the preparation of the bacterial suspension, a liquid yeast-extract medium was prepared. The composition of this liquid medium was identical to the previous solid medium, less the agar. About 10 colonies were extracted from the yeast-extract medium and placed in 100 mL of the liquid medium. Then, a 800 rpm agitation was performed with a magnetic bar at 20 °C ± 1 °C during 60 h. After this period, the absorbance at 600 nm was a bit more than 1 and samples of 50 ml were centrifuged at 5 000 rpm for 15 min to produce, after removing the supernatant, a single pellet of bacteria.

The bacterial solution consisted of 2 pellets in a urea medium to reach an absorbance of about 2.5. The urea medium was composed of 6 g/L Bacto nutrient powder, 40 g/L of Urea, 10 g/L ammonium chloride, 4.24 g/L sodium carbonate at a pH 6 before autoclave.

The fixation solution uses the same urea medium with the addition of equimolar content of calcium chloride. Dehydrated calcium chloride was used, which brings the concentration to 98 g/L.

Finally, the cementation solution was identical to the fixation solution, but the name was different to help in identifying the stage of the protocol.

2.3   Treatment Protocols

Two protocols were successfully elaborated, their difference lying in the curing time. The first step of each protocol was the packing of the dry sand in an Acrylonitrile Butadiene Styrene (ABS) mold of 2 in. (50.8 mm) in diameter and 4 in. (101.6 mm) in height with the methodology to reach the minimum density allowed by the soil, the density is then verified with the known volume of the mold and the weight of the sand. To support the sand, a filter and a perforated cap were placed at the bottom of the mold. The perforated cap was dedicated to the evacuation of the solution in excess during the injection.

The sand was then extracted and mixed by hand with the bacterial solution in a quantity equal to 35% of the void volume. The sand and bacteria were then reintroduced in the mold and a percolation of an identical volume of the fixation solution was started at a flow rate of 120 ml/h from the top of the sample. A curing time of 24 h was allowed to fix the bacteria on the grain particles.

Finally, the cementation phase started. In the first protocol, there were 5 cementation phases separated by 24-h curing periods. For each cementation phase, a volume corresponding to about 110% of the void volume was injected. In the second protocol, only one cementation phase of a volume equivalent of the total volume of the first protocol was performed. In other words, the second protocol was identical to the first one, without the 24-h curing periods.

2.4   Microscopy Observation and X-Ray Analysis

To visualize how the bridges were formed, some sub-samples were taken from the sample treated with the second protocol and analyzed through Scanning Electronic Microscope (SEM) with a JEOL JSM840 apparatus. The samples used to understand the composition thanks to X-ray tests were the same as for the microscopy. They were analyzed with a Philips X'PERT apparatus.

2.5   Calcite Content

The easiest and fastest way to measure the calcite content is to weight the mold and the dry sand before treatment and to compare it with the weight of dry samples after treatment. This procedure was performed on samples dried at 50 °C after treatment. The main advantage of this procedure is that it is easy to implement and reproductible.

2.6   UCS Tests

Unconfined compressive strength tests were performed according to the D-2166 ASTM Standard. After treatment, the samples were flattened using a file and the height and diameter were measured. The speed of the press was adjusted between 0.5% and 2% of the height of the samples.

3   Results and Discussion

3.1   Strength and Calcium Carbonate Content

The results of the compressive tests were put in perspective with the calcium carbonate content in Fig. 1. In this figure, P1 correspond to the first protocol and the P2 to the second one. The duplicates demonstrated the good reproducibility of both protocols. The results show that the Young modulus is largely influenced by the presence of a rest period: mean value equals to 39.70 MPa for P1 and only 11.28 MPa for P2. This observation is mainly due to the low concentration in calcite obtained for P2 specimens, which illustrates the lowest efficiency when there is no rest period during injection steps. As a consequence, the rest periods in protocol P1 produce two times more of calcite and increases the Young modulus by a factor of 3.5 compared to P2.

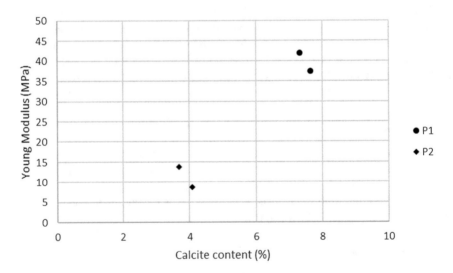

Fig. 1. Young modulus as a function of calcite content

3.2 X-Ray Analysis

An X-ray analysis was performed on a specimen of each protocol and the results are provided in Figs. 2 and 3, respectively for protocols P1 and P2. For protocol P1, it is possible to observe that the measurement is very close to the calcite reference. On the contrary, the comparison to vaterite and aragonite references showed the absence of aragonite and some traces of vaterite. The observation of the P2 specimen offers similar results, the difference being that there is no vaterite in this second sample. Moreover, the presence of silica is self-explanatory as soil grains are 100% pure silica sand. Finally, the investigation presents ammonium chloride as a byproduct of the reaction. As the sample needs to be dried before testing, some residue might still be imprisoned in the water and slowly crystalizing as the water evaporates. Nevertheless, their solubility index is about 414 g/L at 30 °C whereas the calcite is 8.6 10^{-2} g/L at the same temperature (Stchouzkoy-Muxart 1971). The calcite being less soluble than the ammonium chloride, the hypothesis that the augmentation of mass during the process is mainly due to the calcite is plausible. This analysis leaded to the conclusion that the precipitated crystal, cause of the improvement of strength, is calcite, and not vaterite or aragonite.

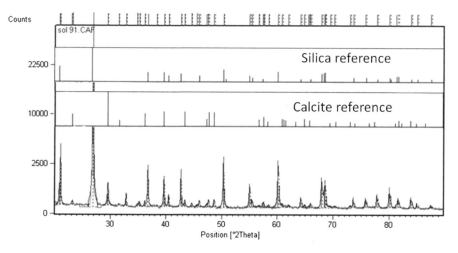

Fig. 2. X-ray analysis (Protocol P1)

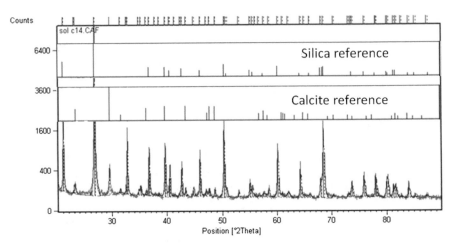

Fig. 3. X-ray analysis (Protocol P2)

3.3   Microscopy Imagery

An example of image taken on P2 specimen is presented in Fig. 4. This sample had a local percentage of calcite of about 3%. The bridges of calcite are visible, and it is

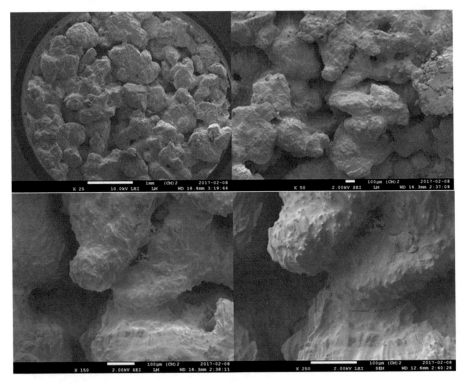

Fig. 4. Microscopic imagery (Protocol P2)

possible to observe a breach in one of them (white arrow in Fig. 4). After the breach of the most fragile zone, the calcite precipitated around the grains is supposed to improve the internal angle friction of the sample, compared to non-treated soils.

4   Conclusions

The goal of this paper was to establish an easy to use protocol to implement MICP in non-saturated conditions. Two protocols were created, and they allowed the samples to reach calcium carbonate content between 3 and 6%. The unconfined compression tests performed on the samples showed that rests periods during injection steps lead to stiffer specimens as well as higher calcite contents. These results are particularly promising for the treatment of loose sand in superficial layers.

Acknowledgments. The authors would like to acknowledge the financial support of the Natural Sciences and Engineering Research Council of Canada (NSERC) and the technical support for bacterial medium preparation provided by the Research, Development and Validation Center for Water Treatment Technology (CREDEAU).

References

Cheng, L., Cord-Ruwisch, R.: Upscaling effects of soil improvement by microbially induced calcite precipitation by surface percolation. Geomicrobiol J. **31**(5), 396–406 (2014)

Cheng, L., Cord-Ruwisch, R., Shahin, M.A.: Cementation of sand soil by microbially induced calcite precipitation at various degrees of saturation. Can. Geotech. J. **50**(1), 81–90 (2013)

Dejong, J.T., Soga, K., Kavazanjian, E., Burns, S.E., Van Paassen, L.A., Al Qabany, A., Aydilek, A., Bang, S.S., Burbank, M., Caslake, L.F., Chen, C.Y., Cheng, X., Chu, J., Ciurli, S., Esnault-Filet, A., Fauriel, S., Hamdan, N., Hata, T., Inagaki, Y., Jefferis, S., Kuo, M., Laloui, L., Larrahondo, J., Manning, D.A.C., Martinez, B., Montoya, B.M., Nelson, D.C., Palomino, A., Renforth, P., Santamarina, J.C., Seagren, E.A., Tanyu, B., Tsesarsky, M., Weaver, T.: Biogeochemical processes and geotechnical applications: progress, opportunities and challenges. Geotechnique **63**(4), 287–301 (2013)

Girinski, O.: Pré-industrialisation d'un procédé de consolidation de sol par bio-calcification in situ. Thèse de l'Université d'Angers (2009)

Hammes, F., Verstraete, W.: Key roles of pH and calcium metabolism in microbial carbonate precipitation. Rev. Environ. Sci. Biotechnol. **1**(1), 3–7 (2002)

Harkes, M.P., van Paassen, L.A., Booster, J.L., Whiffin, V.S., van Loosdrecht, M.C.M.: Fixation and distribution of bacterial activity in sand to induce carbonate precipitation for ground reinforcement. Ecol. Eng. **36**(2), 112–117 (2010)

Martinez, B.C., DeJong, J.T.: Bio-mediated soil improvement: load transfer mechanisms at the micro-and macro-scales. In: Advances in Ground Improvement: Research to Practice in the United States and China - 2009 US-China Workshop on Ground Improvement Technologies, Orlando, FL, United states, American Society of Civil Engineers, 14 March 2009 (2009)

Martinez, B.C., DeJong, J.T., Ginn, T.R., Montoya, B.M., Barkouki, T.H., Hunt, C., Tanyu, B., Major, D.: Experimental optimization of microbial-induced carbonate precipitation for soil improvement. J. Geotech. Geoenviron. Eng. **139**(4), 587–598 (2013)

Montoya, B.M., Dejong, J.T., Boulanger, R.W.: Dynamic response of liquefiable sand improved by microbial-induced calcite precipitation. Geotechnique **63**(4), 302–312 (2013)

Montoya, B.M., Gerhard, R., Boulanger, R.W., Ganchenko, A., Chou, J.-C., DeJong, J.T., Wilson, D.W.: Liquefaction mitigation using microbial induced calcite precipitation. In: GeoCongress 2012 (2012)

Stchouzkoy-Muxart, T.: Contribution à l'étude de la solubilité de la calcite dans l'eau en présence d'anhydride carbonique, à 20°C et 30°C. Bulletin de l'Association de géographes français, pp. 215–226 (1971)

Stocks-Fischer, S., Galinat, J.K., Bang, S.S.: Microbiological precipitation of CaCO3. Soil Biol. Biochem. **31**(11), 1563–1571 (1999)

van Paassen, L.A., Ghose, R., van der Linden, T.J., van der Star, W.R., van Loosdrecht, M.C.: Quantifying biomediated ground improvement by ureolysis: large-scale biogrout experiment. J. Geotech. Geoenviron. Eng. **136**(12), 1721–1728 (2010)

Experimental Investigations on Expansive Soils Grouted with Additives

Tummala Sri Rambabu[1]([✉]), V. V. N. Prabhakara Rao[1],
K. S. R. Prasad[1], and Paladugu Indraja[2]

[1] Civil Engineering, V R Siddhartha Engineering College,
Vijayawada, AP, India
srirambabutummala@yahoo.in, vvnpjee@gmail.com
[2] V R Siddhartha Engineering College, Vijayawada, AP, India
indraja.paladugu@gmail.com

Abstract. Various stabilization techniques have been in practice to counteract the swell-shrink problems posed by expansive soils. Apart from the techniques such as belled piers and under-reamed piles, chemical stabilization of expansive soils has met with considerable success. Various additives used in chemical stabilization. This experimental investigation presents the efficacy of an innovative and comparative study of stabilization technique in the form of columns Fly Ash, lime and GGBS embedded in expansive clay beds. In order to assess the intensity of expansion, swell pressure test is performed on expansive clay beds into which additives are injected which serve as compacted columns. It is observed that swelling decreased significantly in expansive soil used with grouted columns.

Keywords: Expansive soils · Swell pressure · Fly ash · GGBS · Lime · Grouted columns

1 Introduction

Natural expansive soils are very common in India and all over the world. These soils occupy about 30 to 40% of the land area of India. Expansive soil have their own characteristics due to the presence of swelling clay minerals. As they get wet, the clay minerals absorb water molecules and expand; conversely, as they dry they shrink, leaving large voids in the soil. Soils with clay minerals, such as montmorillonite, exhibit the most profound swelling properties. Annually large amount of money is spent to repair the damages caused to infrastructures built on expensive soils. It contains high percentage of clay mineral like montmorillonite and smectite. These soils swell considerably and exhibit low shear strength in dry condition. The seasonal moisture variations in such soils create major damages to the structures. Expansive soils are clays of high plasticity and it exhibits low shear strength. These soils are highly compressible and they contain low bearing capacity. Several treatment methods are currently available for stabilizing the expansive soils, such as mechanical stabilization, chemical stabilization, electrical stabilization, stabilization by geo synthetics, reinforced earth walls, and stabilization using bio enzymes etc.

© Springer Nature Switzerland AG 2020
L. Hoyos and H. Shehata (Eds.): GeoMEast 2019, SUCI, pp. 127–141, 2020.
https://doi.org/10.1007/978-3-030-34206-7_10

2 Literature Review

Abiodun (2015) conducted experimental analyses through clay–lime physicochemical reactions resulting from cation exchange using the lime pile technique and indicated that these reactions have remarkable effects on the electrical properties of the lime pile–treated soil and produced strong inter particle bonds and unconfined compressive strength of the soil. Aljorany et al. (2014) conducted laboratory tests on Granular Pile Anchor (GPA) is embedded in expansive soil, analyzed the whole system using the finite element package PLAXIS software and reported that the heave can be reduced by up to (38%) when GPA is embedded in expansive soil layer at (L/H = 1), while heave was reduced by about (90%) when GPA is embedded in expansive soil. Ashok and Reddy (2016) studied geotechnical improvement of properties of clay soils using the lime pile technique on a laboratory scale model, examined the clay-lime physico-chemical reactions resulting from cation exchange and indicated that these reactions have produced strong inter particle bonds and unconfined compressive strength of the soil. Chandran and Soman (2016) conducted a parametric study on lime mixed GGBS columns in the expansive clay bed formed in a test tank and showed that, the method can be effectively used for reducing heaving problem hence to provide additional strength in an economical manner compared to other methods. Cokca (2001) studied the effect of mixing high-calcium and low-calcium class fly ashes, lime and cement on expansive soil at 0–8% to establish baseline values and showed that addition of 20% fly ash decreased the swelling potential to the swelling potential obtained with 8% lime addition. Darikandeh (2017) conducted laboratory odometer test on expansive soil stabilized by calcium carbide residue–fly ash (CCR–FA) columns with different per-centages of CCR and FA and reported swell potential reduction by 62% and swell pressure reduction by 68% for CCR:FA = 20:80, respectively, the swell pressure and swell potential decreased for all combinations of the CCR–FA admixture, the optimal CCR–FA dosage was found to be 20:80. Dave et al. (2018) performed model plate load tests on expansive clay beds in which fly ash was introduced as compacted columns. The fly ash columns were kept floating 100 mm above the base plate and their results showed that Load-settlement characteristics of expansive clay improved when fly ash was introduced in the form of columns. Dutta and Mandal (2017) studied the use of postconsumer waste plastic water bottles for developing a new type of encasement to confine fly-ash columns fully penetrated in soft clay and observed that, with the increase in mattress height over the encased fly-ash column, the contribution from the encased column decreased accompanying the higher contribution from the geo cell mattress in the overall footing capacity over the encased fly-ash column–geo cell composite systems. Hewayde et al. (2005) investigated the optimum ratio of lime to soil weight to maximize the efficiency of lime columns to reduce the swelling of expansive soil for further enhancing the efficiency of the lime columns reducing soil swelling and showed that the optimum lime content is 6% (by weight) of the soil within the column. Joseph et al. (2018) conducted laboratory studies to determine the improvement of engineering properties of Cochin marine clay by introduction of compacted lime column and lime fly ash column techniques with and without preloading, and indicated that the compacted lime column and lime fly ash column improve the physical, chemical and engineering characteristics of soft clay very effectively. Malekpoor and Poorebrahim (2014) performed laboratory model tests to investigate the behavior of Compacted Lime–Soil (CL-S) rigid stone columns in soft soils. The unit cell idealization is used for construction of composite specimens to evaluate the influence of different parameters such as the diameter of the column (D), the slenderness ratio (L/D) and the area ratio (Ar) and showed that, the load carrying capacity decreases by increasing the slenderness ratio and this ratio has significant influence on the behavior of end bearing columns. More et al. (2017) developed technique in which fly-ash column (FAC) is reinforced with lime and encased by non-woven geo-textile (Polyester) in black cotton soil (BC) and reported that the strength carrying capacity of BC soil was improved and swell-shrink properties were considerably reduced when BC soil was introduced with FAC reinforced with NWG. Muntohar (2010) studied the application of lime-column technique on soft clay soil, along with the strength distribution surround the installed lime-column and the load-settlement characteristic in laboratory and observed that soil strength tends to increase with time and reported that the bearing capacity of the soft soil increased from 0.23 kN to 5.2 kN after the lime column was installed. Phanikumar et al. (2009) conducted heave tests and compressive load tests on expansive clay beds into which fly ash was introduced as compacted columns and reported that heave as well as stress-settlement characteristics of expansive clay beds decreased significantly on reinforcing the clay beds with FACs. Porbaha and Hanzawa (2001) studied the effectiveness of vertical fly ash columns as a technique for the modification of soft-ground properties using Mikasa's apparatus and demonstrated that stress-displacement curves of composite samples of clay-sand and clay-fly ash increase gradually and reach a plateau at small strain levels. Sahoo et al. (2008) found that the red soil cushion over expansive soil is beneficial to reduce its swell-shrink potentials with cycles of wetting and drying only under nominal surcharge, but it is seen to be effective only in the first cycle of swelling, for the case with higher surcharges. Sharma and Sivapullaiah (2012) reported the findings of laboratory tests carried out on local Indian expansive black cotton soil with GGBS mixed with the expansive soil in different proportions and shown that the improvement in initial tangent modulus with GGBS content is very high up to 20% of GGBS addition but beyond this content the change is very small. Shen et al. (2003) reported that clay fracturing is the basic factor involved in property changes and strength increases in clay surrounding DM columns. Wilkinson et al. (2010) reported that fluid pressure induced within the soil mass by the injection process affects the form of observed lime/fly ash deposits entrained within the soil profile. Yadu and Tripathi (2013) reported that OMC increased and MDD decreased with the addition of fly ash-granulated blast furnace slag mixture to the soft soil in both soaked and unsoaked condition.

   Review of above literature discloses that, there are field/ laboratory trials using powders of fly ash, GGBS and lime either directly mixed with expansive soil or as a column of material impregnated into expansive clay. In addition, lime columns are used to study cation exchange capacity. Numerical studies are made, marginal soils are used as soil cushion. But, a physical model to compare suitability of selected few additives in modeled column of additives requires attention. Our study focused on this.

3 Methodology

3.1 Expansive Soil

Soil samples are collected from different regions namely Rudravaram (100% free swell index-soil[1])) and Ibrahimpatnam (80% free swell index-soil[2]) near Vijayawada at 1 m depth. Collected samples are oven dried at temperature of 105–110 °C and are pulverized. Basic properties are determined as per IS-1498 & IS-2720. The physical properties of Expansive clayey soil used in this research work are given in Table 1. Standard proctor compaction test was conducted to know optimum moisture content (OMC) and maximum dry density (MDD) values. Swell Pressure intensity is calculated from proving ring reading and specimen area. A pressure versus time graph is plotted. The maximum pressure intensity gives the swelling pressure of soil for a specific dry density and water content. Constant pressure method is adopted and load intensities are so selected that soil swell under lowest load intensity and consolidate under maximum load intensity. After the equilibrium is achieved, the changes in the volume of specimen are recorded. A graph between load intensity as abscissa and volume change as ordinate. The load intensity at which volume change is zero is called swell pressure.

Table 1. Physical properties of natural soil samples

S. No	Description	soil[1]	soil[2]
1	Gravel (%)	0	0.8
2	Sand (%)	7.7	11.8
3	Silt and clay (%)	92.3	87.4
4	Specific gravity	2.62	2.65
5	Liquid limit (%)	78	80
6	Plastic limit (%)	39.58	26.43
7	Shrinkage limit (%)	54.50567	56.743
8	Free swell index (%)	100	80
9	Max dry density (kN/m^3)	14.08716	14.1264
10	Optimum moisture content (%)	27.5	28
11	**Swell pressure (kg/sq.cm)**	**1.0152295**	**0.82109**
12	BIS soil classification	CH	CH

3.2 Grout Materials

Fly ash for the present study is collected from Narla TataRao Thermal power Station, Ibrahimpatnam (Near Vijayawada) A.P. India. GGBS is collected from Visakhapatnam steel plant, Visakhapatnam, A.P. India. Lime is procured from Merck Specialties Private Limited, Vijayawada, A.P. India. The basic physical properties of the Fly ash and GGBS are listed in Table 2. The chemical properties of the Fly ash and GGBS are listed in Table 3. Specification of lime is listed in Table 4.

Table 2. Physical properties of fly ash and GGBS

S. No	Description	Fly ash	GGBS
1	Specific gravity, Gs	2.16	3
	Particle size distribution (%)		
2	Gravel (>4.75 mm)	0	0
3	Sand (0.74 to 4.75 mm)	23	1
4	Silt (0.002 to 0.74 mm)	15	36
5	Clay (<0.002 mm)	62	63
6	Maximum dry density, (kN/m3)	13.4	16.5
7	Optimum moisture content, (%)	18.76	19.3

Table 3. Chemical properties of fly ash and GGBS

S. No	Composition	Fly ash	GGBS
1	$Si\,O_2$	57.2	30
2	$Al_2O_3\%$	24.1	16
3	$Fe_2O_3\%$	4.6	4.5
4	CaO %	1.5	35
5	MgO	3	9
6	Mn	0.3	1.5
7	Other chemicals	8.5	3.5
8	Loss on ignition	0.8	0.5

Table 4. Chemical composition of lime

S. No	Composition	Lime
1	Calcium oxide	≥ 90
2	Chloride	Passes test
3	Loss on ignition	≤ 10

3.3   Experimental Investigations

Soil is initially compacted at Optimum Moisture Content (OMC), Maximum Dry Density (MDD) and swell pressure test is conducted on the compacted soil samples. Groove is made in the soil samples with 0.6 mm diameter and in a grid pattern. The number of columns gradually increased from 4 to 10 (Increment = 2 columns in every trial). Slurry of additives i.e. GGBS, fly ash is prepared with ratio of additives to water as 1:1. Slurry is injected into these columns and specimen left for maturing for 7 days and swell pressure is conducted on these specimens as shown in Fig. 1.

Fig. 1. 6 mm diameter columns grouted with industrial waste slurry in swell pressure mould

4   Results

Tables 5 and 8 shows the variation of swell pressure with different industrial waste grouted material for soil[1] and soil[2] respectively, for lime columns 100% swell pressure reduction took place at 10 columns, for GGBS it is at 12 columns and for Fly ash 100% swell pressure reduction is achieved at 16 columns (Figs. 2 and 5).

Table 5. Swell pressure variation of different grout materials for soil[1]

No of columns	Lime columns swell pressure (kg/sq.cm)	GGBS columns swell pressure (kg/sq.cm)	Fly ash columns swell pressure (kg/sq.cm)
0	1.01522	1.01522	1.01522
4	0.76615	0.792	0.9153
6	0.5916	0.625	0.7924
8	0.4167	0.47387	0.6518
10	0	0.2034	0.5031
12		0	0.43702
14			0.3008
16			0

Table 6. Percentage reduction of swell pressure for different grout materials for soil[1]

No of columns	% Reduction of swell pressure of lime columns	% Reduction of swell pressure of GGBS columns	% Reduction of swell pressure of Fly ash columns
0	0	0	0
4	24.5336	21.9874	9.8423
6	41.727	38.437	21.948
8	58.9548	53.3235	35.7972
10	100	79.965	50.4443
12		100	56.9532
14			70.371
16			100

Table 7. Volume of grout required for stabilizing the soil[1]

No of columns	Volume of lime grout injected (CC)	Volume of GGBS grout injected (CC)	Volume of Fly ash grout injected (CC)
0	0	0	0
4	56.56	56.56	56.56
6	84.84	84.84	84.84
8	113.12	113.12	113.12
10	141.4	141.4	141.4
12		169.68	169.68
14			197.96
16			226.24

Table 8. Swell pressure variation of different grout materials for soil[2]

No of columns	Lime columns swell pressure (kg/sq cm)	GGBS columns swell pressure (kg/sq cm)	Fly ash columns swell pressure (kg/sq cm)
0	0.82109	0.82109	0.82109
4	0.57191	0.661	0.759
6	0.4085	0.47795	0.66
8	0.2139	0.286	0.6
10	0	0	0.502
12			0.4
14			0.25736
16			0

Tables 6 and 9 shows the percentage reduction of swell pressure or different industrial waste grouted column for soil[1] and soil[2] respectively, for lime columns 100% swell pressure reduction took place at 10 columns, for GGBS it is at 12 columns and for Fly ash 100% swell pressure reduction is achieved at 16 columns (Figs. 3 and 6).

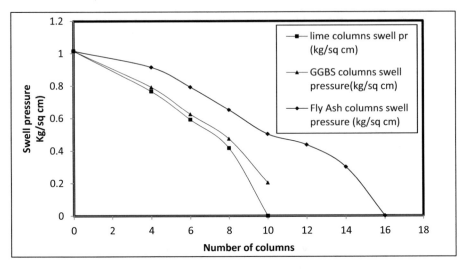

Fig. 2. Swell pressure v/s Number of columns for soil[1]

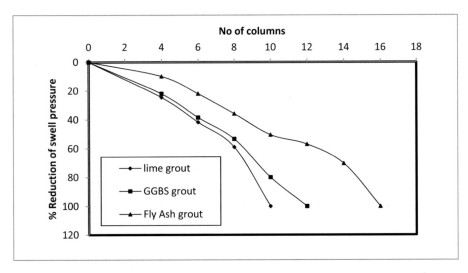

Fig. 3. Percentage reduction of swell pressure v/s Number of columns for soil[1]

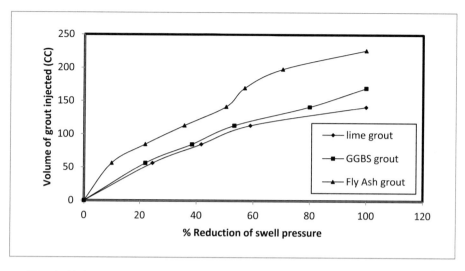

Fig. 4. Volume of grout injected v/s Percentage reduction of swell pressure for soil[1]

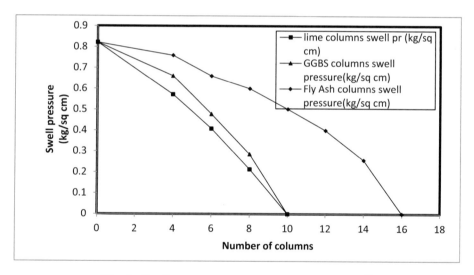

Fig. 5. Swell pressure v/s Number of columns for soil[2]

Tables 7 and 10 shows that, the volume required to minimize the swell pressure by 100% for soil[1] and soil[2] respectively, for lime columns 141.4CC is required for 10 columns to reduce the heave, for GGBS 169.98CC is required for 12 columns to reduce the heave and for Fly ash 226.24CC is required for 16 columns to reduce the heave by 100% (Figs. 4 and 7).

Table 9. Percentage reduction of swell pressure for different grout materials for soil[2]

No of columns	% Reduction of swell pressure of Lime columns	% Reduction of swell pressure of GGBS columns	% Reduction of swell pressure of Fly ash columns
0	0	0	0
4	30.3475	19.4973	7.5619
6	50.2491	41.7908	19.6191
8	73.9493	65.1683	26.9265
10	100	100	38.8618
12			51.2843
14			68.6563
16			100

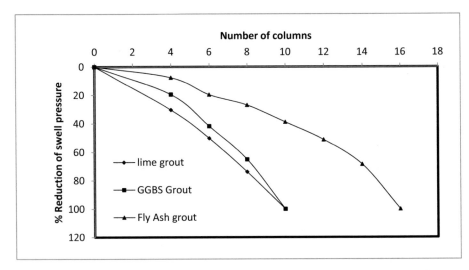

Fig. 6. Percentage reduction of swell pressure v/s Number of columns for soil[2]

Table 10. Volume of grout required for stabilizing the soil[2]

No of columns	Volume of Lime grout injected (CC)	Volume of GGBS grout injected (CC)	Volume of Fly ash grout injected (CC)
0	0	0	0
4	56.56	56.56	56.56
6	84.84	84.84	84.84
8	113.12	113.12	113.12
10	141.4	141.4	141.4
12			169.68
14			197.96
16			226.24

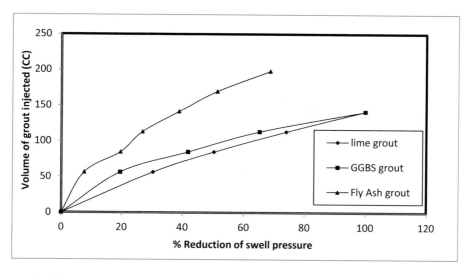

Fig. 7. Volume of grout injected v/s Percentage reduction of swell pressure for soil²

Tables 11, 12 and 13 shows the swell pressure reduction for the soil¹ and soil²
using Lime, GGBS and Fly Ash grouted columns respectively. It is observed that 10
lime columns are required to reduce the swell pressure by 100% for soil¹ and soil².
For GGBS grouted columns swell pressure reduction of 100% is achieved at 12 columns and 10 columns for soil¹ and soil² respectively. For Fly ash columns it is
achieved at 16 columns for both the soils (Figs. 8, 9 and 10).

Table 11. Swell pressure reduction for different soils using lime grouted column

No of columns	soil¹	soil²
0	1.01522	0.82109
4	0.76615	0.57191
6	0.5916	0.4085
8	0.4167	0.2139
10	0	0

Table 12. Swell pressure reduction for different soils using GGBS grouted column

No of columns	soil¹	soil²
0	1.01522	0.82109
4	0.792	0.661
6	0.625	0.47795
8	0.47387	0.286
10	0.2034	0
12	0	

Table 13. Swell pressure reduction for different soils using Fly ash grouted column

No of columns	soil[1]	soil[2]
0	1.01522	0.82109
4	0.9153	0.759
6	0.7924	0.66
8	0.6518	0.6
10	0.5031	0.502
12	0.43702	0.4
14	0.3008	0.25736
16	0	0

Swellpressure comparison of soil1& soil2: (When used with different grouts)

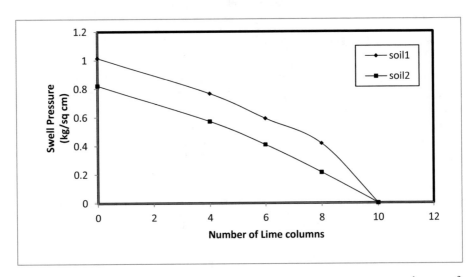

Fig. 8. Swell Pressure v/s Number of columns for Lime grouted columns for soil[1] and soil[2]

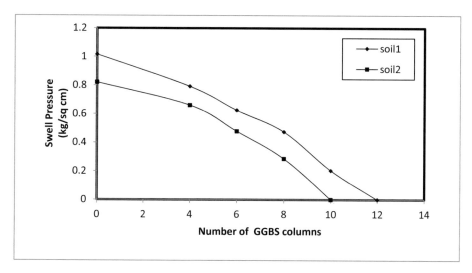

Fig. 9. Swell Pressure v/s Number of columns for GGBS grouted columns for soil[1] and soil[2]

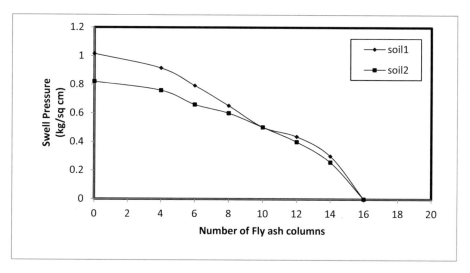

Fig. 10. Swell Pressure v/s Number of columns for Fly Ash grouted columns for soil[1] and soil[2]

5   Conclusions

From the experimental investigations the following observations can be drawn

1. When lime is used as stabilizer, swell pressure is reduced from 0.82109 kg/cm^2 to 0 kg/cm^2, with varying number of lime piles with an increment of 2, from 4 to 10 respectively, with diameter of 6 mm.

2. Swell pressure is reduced from 0.82109 kg/cm^2 to 0 kg/cm^2, when GGBS is used as stabilizer with varying number of lime piles with an increment of 2, from 4 to 10 respectively, with diameter of 6 mm.
3. Fly ash also played a key role in minimizing the swell pressure from 0.82109 kg/cm^2 to 0 kg/cm^2, with varying number of columns from 4 to 16 for both the soils.
4. Thus, Lime and GGBS are more effective in minimizing the swell pressure, compared to fly ash.

References

Abiodun, A.A., Nalbantoglu, Z.: Lime pile techniques for the improvement of clay soils. Can. Geotech. J. (2015). https://doi.org/10.1139/cgj-2014-0073

Aljorany, A.N., et al.: Heave behavior of granular pile anchor-foundation system. J. Eng. (2014). https://www.iasj.net/iasj?func=fulltext&aId=87355

Ashok, P., Reddy, G.S.: Lime pile technique for the improvement of properties of clay soil. Int. J. Sci. Res. (IJSR) (2016). https://pdfs.semanticscholar.org/0382/8d043e6f7355be313c43d811e04ea7cde374.pdf

Chandran, G., Soman, K.: Heave control and strengthening of expansive soil using lime mixed GGBS column. Int. J. Eng. Res. Technol. (IJERT) (2016). https://www.ijert.org/research/heave-control-and-strengthening-of-expansive-soil-using-lime-mixed-ggbs-column-IJERTV5IS080345.pdf

Cokca, E.: Use of class c fly ashes for the stabilization of an expansive soil. J. Geotech. Geo Environ. Eng. (2001). https://doi.org/10.1061/(ASCE)1090-0241(2001)127:7(568)

Darikandeh, F.: Expansive soil stabilized by calcium carbide residue–fly ash columns. Proc. Inst. Civ. Eng. Ground Improv. (2017). https://doi.org/10.1680/jgrim.17.00033

Dave, S.P., et al.: Load-settlement characteristics of expansive soils treated with fly ash columns. IJARIIE-ISSN(O)-2395-439 (2018). http://ijariie.com/AdminUploadPdf/LOAD_SETTLEMENT_CHARACTERISTICS_OF_EXPANSIVE_SOILS_TREATED_WITH_FLY_ASH_COLUMNS_ijariie8256.pdf

Dutta, S., Mandal, J.N.: Model studies on encased fly ash column–geocell composite systems in soft clay. J. Hazard. Toxic Radioactive Waste (2017). https://doi.org/10.1061/(ASCE)HZ.2153-5515.0000353

Hewayde, E., et al.: Reinforced lime columns: a new technique for heave control. Proc. Inst. Civ. Eng. Ground Improv. (2005). https://doi.org/10.1680/grim.2005.9.2.79

Joseph, A., et al.: Performance of compacted lime column and lime-fly ash column techniques for cochin marine clays. Int. J. Geosynth. Ground Eng. (2018). https://doi.org/10.1007/s40891-018-0147-5

Malekpoor, M.R., Poorebrahim, G.R.: Behavior of compacted lime-soil columns. Int. J. Eng. (2014). https://doi.org/10.5829/idosi.ije.2014.27.02b.15

More, S., et al.: Laboratory study on soil reinforced with fly ash columns with and without encasement of non woven geotextile. IRJET (2017). https://link.springer.com/article/10.1007/s12665-018-7287-8

Muntohar, A.S.: A laboratory test on the strength and load-settlement characteristic of improved soft soil using lime-column. Terbitan Berkala Ilmiah J. (2010). http://hdl.handle.net/11617/1693

Phanikumar, B.R., et al.: Fly ash columns (FAC) as an innovative foundation technique for expansive clay beds. Geo Mech. Geo Eng. Int. J. (2009). https://doi.org/10.1080/17486020902857007

Porbaha, A., Hanzawa, H.: Ground modification by vertical fly ash columns. Proc. Inst. Civ. Eng. Ground Improv. (2001). https://doi.org/10.1680/grim.2001.5.3.101

Sahoo, J., et al.: Behavior of stabilized soil cushions under cyclic wetting and drying. Int. J. Geotech. Eng. (2008). https://doi.org/10.3328/IJGE.2008.02.02.89-102

Sharma, A.K., Sivapullaiah, P.V.: Improvement of strength of expansive soil with waste granulated blast furnace slag. Geo Congress (2012). https://doi.org/10.1061/9780784412121.402

Shen, S.L., et al.: Interaction mechanism between deep mixing column and surrounding clay during installation. Can. Geotech. J. (2003). https://doi.org/10.1139/t02-109

Wilkinson, A., et al.: Improvement of problematic soils by lime slurry pressure injection: case study. J. Geotech. Geo Environ. Eng. (2010). https://doi.org/10.1061/(ASCE)GT.1943-5606.0000359

Yadu, L., Tripathi, R.K.: Stabilization of soft soil with granulated blast furnace slag and fly ash. Int. J. Res. Eng. Technol. (2013). https://doi.org/10.15623/ijret.2013.020200

Multivariate Regression Analysis in Modelling Geotechnical Properties of Soils Along Lambata-Minna-Bida Highway

S. H. Waziri[1(✉)], F. Attah[2], and N. M. Waziri[1]

[1] Department of Geology, Federal University of Technology,
Minna, Niger State, Nigeria
salwaz1969@gmail.com
[2] Road Research Department, Nigerian Building and Road Research Institute,
km 10 Idiroko Road, Otta, Ogun State, Nigeria

Abstract. In pavement design, three important geotechnical properties – CBR, OMC, and MDD are often used to determine the strength of a subgrade layer. To determine these properties in the laboratory is time consuming, laborious, very costly and sometimes infrequently performed due to lack of equipment. The aim of this study is therefore to develop regression models to estimate the strength properties using relatively easier index properties. Thirty – four soil samples were collected from various locations along Bida – Minna highway between 0.6–1.5 m depths for index, consistency, compaction and CBR tests. Based on the laboratory results, the CBR significantly related with sand, % fines, LL, PL, PI, OMC and MDD parameters. Satisfactory empirical correlations ($R^2 > 0.59$) were found between the three strength properties and other index properties of the experimented soils. Seven best predictive models were developed to estimate the strength properties based on multiple linear regression analysis.

Keywords: California bearing ratio · Soil physical properties · Pavement · Multiple regression analysis

1 Introduction

Roads are major means of transportation and economic development and most of them consist of flexible pavement. Flexible pavement is made up of different layers such as subgrade, subbase, base course and surface layer. Design and performance of flexible pavement mainly depends on the strength of subgrade material. The load from the pavement surface is ultimately transferred to subgrade and to the sub-base. The subgrade is designed such that the stress transferred should not exceed elastic limit. Hence, the suitability and stability of subgrade material is evaluated before construction of pavement. Soaked California bearing ratio (CBRs) value is considered a major strength parameter in subgrade design that influences the thickness of the subgrade. If the CBR value is higher, then the designed thickness of the subgrade is thinner and vice versa. Technically, CBR test is difficult to determine in a short duration, very expensive, tedious and requires large quantity of soil sample and consequently, leads to delay and high cost implication on any structural project. In Nigeria, most of these paved roads

© Springer Nature Switzerland AG 2020
L. Hoyos and H. Shehata (Eds.): GeoMEast 2019, SUCI, pp. 142–149, 2020.
https://doi.org/10.1007/978-3-030-34206-7_11

and other earth structures have witnessed different degrees of failures, which had become a cause for concern to geoscientists. To overcome this problem, a simple and less time consuming technique becomes inevitable through linear regression analysis of soaked CBR value with other easily determined soil properties.

Over the years, several researchers (Egbe et al. 2017; Noor 2011; Talukdar 2014; Ramasubbarao and Siva 2013; Akshay 2013; Patel and Desai 2010) have applied multiple linear regression analysis (MLRA) in modelling geotechnical soil properties and many correlations were developed. In addition, several attempts were made to predict CBR values based on physical properties such as index properties (Nguyen and Mohajerani 2015; Gregory 2007; Pal and Ghosh 2010). In this paper, soaked CBR is correlated with other soil properties such as liquid limit, plastic limit, and plasticity index, optimum moisture content (OMC), maximum dry density (MDD) and percentage fines using simple and multiple regression analyses. Their findings showed a strong correlation between the CBR and fines as well as plasticity index. Giasi et al. (2003) studied the index properties, such as liquid limit and plasticity index of various soils and proposed a numerous equations.

2 Materials and Methods

2.1 The Study Area and Geologic Setting

The study area is the Lambata – Minna highway, 105 km in distance, and located between longitude 6°20′–6°35′E and latitude 9°05′–9°35′N. The highway is underlain by Precambrian basement complex, made up of biotite granite, granite gneiss, migmatite, marble and schist. The granite has been affected by the Pan African Orogeny with late tectonic emplacement of granites and granodiorites (Olayinka 1992). Different disturbed soil samples were collected from different locations along the highway and subjected to geotechnical laboratory tests according to BS 1377 (1990) and West African standards. Index properties (grain size distribution and consistency limits), compaction (OMC and MDD using Standard Proctor mould) and CBR tests were conducted on the soils samples.

2.2 Relationship Between Soil Properties

To establish relationship between soaked CBR and different soil properties, scatter plots were generated with CBR against different soil parameters and suitable trend line was drawn with higher correlation coefficient. Correlation quantifies the degree to which dependent and independent variables are related. Linear regression quantifies goodness of fit with the coefficient (R^2) value. R^2 value provides a measure of how well future outcomes are likely to be predicted by the model.

2.3 Multiple Linear Regression Analysis (MLRA)

To develop the models of multiple linear regression analysis, soaked CBR value was considered as dependent variable and soil properties such as gravel (G), fines (F), sand (S),

LL, PL, PI, OMC and MDD as the independent variables using SPSS and Microsoft Excel software to derive the relationship statistically. The regression equation is in Eq. 1.

$$Y = b_0 + b_1 x_1 + b_2 x_2 + b_3 x_3 + \ldots + b_n x_n \tag{1}$$

Where

Y is the dependent variable (i.e. CBRs)
$b_1 - b_n$ the intercept (constant), b_0 the slope (regression coefficient)
$x_1 - x_n$ the independent variables (soil properties considered in the analysis).

3 Results and Discussions

The geotechnical properties of thirty-nine soil samples collected are summarized in Table 1. Soils occurring within the granite and grandiorite terrain are well graded soils with high gravels (39.9%, 44.3%), moderate sands (46.4%, 36.7%) and high fines (11.7%, 14.5%) respectively. The sandstone terrain has high presence of sands (80%), MDD (2.0 mg/m³), soaked and unsoaked CBR (76.4% and 112.4%) respectively. The migmatite gneiss terrain has high liquid content (38.2%), high OMC (21.0%), very low CBRs (8.5%).

From Table 1, it is observed that plasticity depends on grain size of soils. When the sand content in the soil increases the plasticity index decreases. Due to the decrease in attraction force, liquid limit of the soil decrease and accordingly plasticity index decrease. With the increase in fines intermolecular attraction force increases resulting in increase in liquid limit. Nath and Dalal (2004) also observed the same trend in their studies.

Table 1. Summary of geotechnical soil test

Proth		G	S	F	LL	PL	PI	OMC	MDD	CBRs	CBRu
GN = 17	MIN	6.4	15.9	2.0	24.0	2.5	12.6	12.2	1.4	2.6	2.7
	MAX	80.7	79.7	47.9	49.0	20.2	42.0	29.9	2.1	115.0	121.8
	MEAN	39.9	46.4	11.7	35.5	7.7	27.6	19.1	1.8	41.5	58.0
	STD	21.0	19.4	10.5	7.1	5.1	9.3	5.2	0.2	24.9	33.9
SS = 8	MIN	5.9	71.3	2.5	14.5	1.0	3.1	9.4	1.7	13.0	74.0
	MAX	25.4	91.6	8.8	45.0	36.9	40.0	18.6	2.2	132.4	144.4
	MEAN	14.9	80.0	5.2	26.6	10.8	16.0	12.1	2.0	76.4	112.4
	STD	5.7	6.1	2.5	11.3	12.1	11.0	2.8	0.1	34.1	24.8

(*continued*)

Table 1. (*continued*)

Proth		G	S	F	LL	PL	PI	OMC	MDD	CBRs	CBRu
MG = 10	MIN	4.3	14.2	2.3	28.7	5.7	13.0	14.1	1.5	0.9	1.4
	MAX	70.2	89.4	15.2	50.1	27.0	30.5	28.3	2.1	24.1	70.6
	MEAN	37.2	52.3	8.1	38.2	17.0	21.2	21.0	1.8	8.5	14.4
	STD	20.4	23.5	3.9	6.0	9.6	6.5	5.0	0.2	7.1	20.5
GD = 2	MIN	21.0	11.6	10.6	35.0	2.5	27.5	21.1	1.6	2.6	3.3
	MAX	67.7	61.9	18.5	40.0	7.5	37.5	23.7	1.9	4.3	4.9
	MEAN	44.3	36.7	14.5	37.5	5.0	32.5	22.4	1.7	3.5	4.1
	STD	33.0	35.6	5.5	3.5	3.5	7.1	1.8	0.2	1.2	1.1
SH = 1		24.1	54.7	21.0	38.5	5.4	23.1	31.7	1.8	18.7	19.6
MI = 1		33.8	50.2	16.0	39.0	8.0	31.0	20.0	1.7	33.4	35.9

G = gravels (%), S = sand (%), F = fines (%), LL = liquid limit (%), PL = plastic limit (%), PI = plasticity index (%), OMC = optimum moisture content (%), MDD = maximum dry density (mg/m^3), CBRs = soaked California bearing ratio (%), CBRu = unsoaked California bearing ratio (%), PROTH = Protholic rocks, GN = Granite, SS = Sandstone, MG = Migmatite gneiss, GD = Granodiorite, SH = Schist, MI = Migmatite

3.1 Relation Between Soaked CBR and Soil Properties

The statistical evaluation of soil properties has always attracted the interest of geotechnical engineers. In this section, various correlations have been established between different soil parameters. The relationship between CBR and different soil properties are developed and mathematical equations are shown in Fig. 1. The variation between soaked CBR and both liquid limit and plasticity index showed a suitable trend line with a third degree polynomial equation. However, the influence of plasticity index on CBRs is not clear as the points scatterred about in the plot. As the moisture content increases, there is progressive reduction in CBR with corresponding decrease in MDD due to the reduction in shear strength and the density of the fine-grained soils. This therefore implies that moisture content influences the strength of soil. This observation agrees with the reports from Nguyen and Mohajerani (2015).

The correlation matrix of different soil properties is shown in Table 2. Relationship of soil properties vary from positive to negative correlation. There is a weak correlation of CBRs and other soil properties (R < 0.5) except for OMC (R = 0.6). Furthermore, the correlation with Atterberg limit indicates that CBR correlate better with liquid limit than plastic limit.

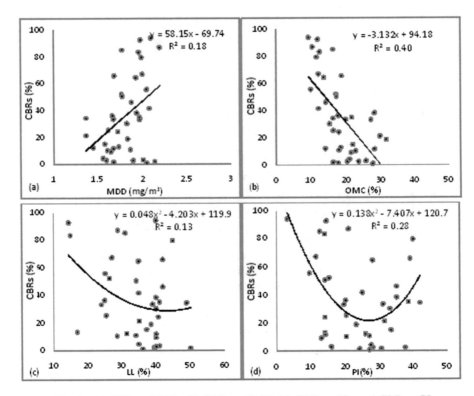

Fig. 1. (a) CBR vs MDD, (b) CBR vs OMC, (c) CBR vs LL, and CBR vs PI

Table 2. Correlation matrix of soil properties

	G	S	F	LL	PL	PI	OMC	MDD	CBRs	CBRu
G	1									
S	−0.91	1								
F	−0.16	−0.16	1							
LL	0.50	−0.49	−0.14	1						
PL	−0.10	0.16	−0.22	0.34	1					
PI	0.53	−0.56	0.04	0.55	−0.59	1				
OMC	0.36	−0.47	0.26	0.23	−0.07	0.22	1			
MDD	−0.14	0.27	−0.25	−0.16	0.07	−0.21	−0.30	1		
CBRs	−0.37	0.46	−0.20	−0.37	−0.03	−0.28	−0.60	0.39	1	
CBRu	−0.36	0.50	−0.29	−0.45	−0.02	−0.36	−0.65	0.41	0.87	1

3.2   Multiple Linear Regression Analysis

Besides correlating the soil parameters in order to examine their relationships, the relationship of strength properties (OMC, MDD and CBRs) with index properties such as grain size distribution and Atterberg limit were also investigated. Several models were constructed and three best fit models were selected and shown in Table 3. The three strength properties were estimated and verified through the standard error and its significance. The errors in the values of CBRs, MDD and OMC obtained from the equations are within the range of +29.9, +0.2 and +5.6 respectively.

From the Table 3, it is observed that the correlation coefficient varies significantly from 0.39 to 0.65 for different soil functions. Regression was also developed for CBRs as a function of moisture content, maximum dry density and plasticity index. It should be noted that in the field, the subgrade soil is recommended to be compacted at the OMC initially in order to achieve the MDD. Under severe changes in seasonal climate and drainage conditions, the moisture content of the subgrade soils increases over the service life of the constructed road. This is buttressed by the significantly higher R^2 value of 0.68 in model no. 3.

Table 3. Regression equation for CBRs, MDD and OMC

Model no.	Regression equation	R^2	Significance	F
1	CBRs = 35.8 − 2.01(LL) + 1.21(PI) + 1.01(PL) + 0.13(G) + 0.61(S) − 0.59(F)	0.53	0.08	2.05
2	CBRs = 40.4 + 0.13(G) + 0.58(S) − 0.63(F) − 0.95(LL)	0.52	0.02	3.19
3	CBRs = 41.85 − 0.41(PI) − 2.88(OMC) + 32.77(MDD)	0.68	0.00	8.45
4	MDD = 1.39 + 0.015(LL) − 0.019(PI) − 0.017(PL) + 0.006(G) + 0.007(S) − 0.002(F)	0.42	0.36	1.13
5	MDD = 1.28 + 0.006(G) + 0.007(S) − 0.002(PI)	0.39	0.21	1.56
6	OMC = 24.12 + 0.94(LL) − 0.95(PI) − 0.91(PL) − 0.006(G) − 0.122(S) + 0.103(F)	0.57	0.03	2.59
7	OMC = 24.91 − 0.003(G) − 0.124(S) + 0.137(F) − 0.031(PI)	0.51	0.03	2.95

To examine the error in the prediction, the predicted strength parameters were plotted against the actual soil strength values in Fig. 2. From the below plots, it can be seen that the variations between the predicted and the measured values are not so significant particularly with OMC and MDD. Hence, can be considered as good prediction models.

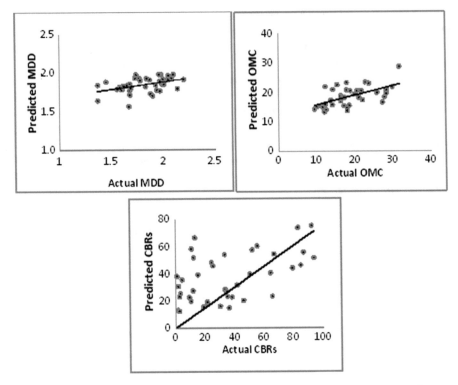

Fig. 2. The predicted strength values versus the experimented values for all soils

4   Conclusion

Maximum dry density and optimum moisture content are important strength properties used for preliminary investigation, design and control. This study investigated the relationship of different soil properties as a function of strength of the road. Multilinear regression analyses were conducted to develop predictive model from statistical point of view to estimate strength parameters such as CBR, MDD, OMC. Several conclusions have been drawn from the study:

  i. With the increase in finer fraction and liquid limit, plasticity index increases.
 ii. The effect of moisture content on CBRs is significant. As the moisture increases, the CBR decreases significantly.
iii. MDD and OMC are best correlated with plasticity index compared to liquid and plastic limits.
 iv. MLRA provides reliable predictive models. However, these models should be limited to soils with similar characteristics and more samples are recommended to improve the prediction.

v. The regression of CBRs and moisture content, plasticity index and maximum dry density was found to be significantly strong ($R^2 = 0.68$).

vi. With less percentage errors, the empirical relations can be accepted as useful information for engineers in the field for preliminary design and estimation.

References

British Standard (BS) 1377: Methods of testing soils for civil engineering purposes. British Standards Institution, London (1990)

Egbe, J.G., Ewa, D.E., Ubi, S.E., Ikwa, G.B., Tumenayo, O.O.: Application of multilinear regression analysis in modelling of soil properties for geotechnical civil engineering works in Calabar South. Niger. J. Technol. **36**(4), 1059–1065 (2017)

Giasi, C.I., Cherubini, C., Paccapelo, F.: Evaluation of compression index of remoulded clays by means of Atterberg limits. Bull. Eng. Geol. Environ. **62**, 333–340 (2003)

Gregory, G.H.: Correlation of California bearing ratio with shear strength parameters. Transp. Res. Rec. J. Transp. Res. Board **1989**, 148–153 (2007)

Nath, A., Dalal, S.S.: The role of plasticity index in predicting compression behavior of clays. Electron. J. Geotech. Eng. **9**, 1–7 (2004)

Nguyen, B.T., Mohajerani, A.: Prediction of California bearing ratio from physical properties of fine-grained soils. Int. J. Civ. Environ. Struct. Constr. Archit. Eng. **9**(2), 1–6 (2015)

Noor, S.C.: Estimation of proctor of compacted fine grained soils from index and physical properties. Int. J. Earth Sci. Eng. **4**, 147–150 (2011)

Olayinka, A.I.: Geophysical siting of boreholes in crystalline basement areas of Africa. J. Afr. Earth Sci. **14**, 197–207 (1992)

Pal, S.K., Ghosh, A.: Influence of physical properties on engineering properties of class F fly ash. In: Indian Geotechnical Conference, December 2010, pp. 361–364. IIT Bombay (2010)

Patel, R.S., Desai, M.D.: CBR predicted by index properties for alluvial soils of South Gujarat. In: Indian Geotechnical Conference, GEO Trend, 16–18 December 2010, pp. 79–82. IGS Mumbai Chapter and IIT Bombay (2010)

Ramasubbarao, G.V., Siva, S.G.: Predicting soaked CBR value of fine grained soils using index and compaction characteristics. Jordan J. Civ. Eng. **7**(3), 354–360 (2013)

Talukdar, D.K.: A study of correlation between California bearing ratio (CBR) value with other properties of soil. Int. J. Emerg. Technol. Adv. Eng. **4**(1), 559–562 (2014)

Prediction of Geotechnical Properties of Lime-Stabilized Soils: Ongoing Research and Preliminary Results

Muhannad Ismeik[✉] and Taha Ahmed

Department of Civil Engineering, Australian College of Kuwait,
Safat, 13015 Kuwait City, Kuwait
{m.ismeik, t.ahmed}@ack.edu.kw

Abstract. This paper presents statistical models to estimate the geotechnical properties of lime-stabilized fine-grained soils. The geotechnical properties of the un-stabilized and lime-stabilized soils are measured based on the American Society for Testing and Materials (ASTM) standards. These properties include liquid limit, plasticity index, optimum moisture content, maximum dry density, and unconfined compressive strength. Findings indicated that the suggested regression equations exhibit excellent fit of data and can be used reliably and efficiently to predict the geotechnical behavior of lime-stabilized soil, as a rapid inexpensive substitute for cumbersome laboratory techniques. These models can be directly implemented into the design and construction of engineering earthworks including road subgrades, landfill liners, and foundations.

Keywords: Modeling · Lime · Stabilization · Fine-grained soil · Subgrade · Foundation · Geotechnical properties

1 Introduction

Index properties of fine-grained soils are very important since they influence the engineering behavior of such soils and they are linked to soil type and mineralogy. These properties include plastic properties, maximum dry density, optimum moisture content, strength, swelling, and consolidation. Measuring these properties is typically carried out experimentally in the laboratory, which usually takes considerable amount of time and effort.

Knowledge of index properties is necessary as to determine the suitability of soil for an engineering application including earth filling and selection of subgrade for pavement design. Very often, the soil is poor and requires some sort of treatment prior to engineering use. Generally, mixing fine-grained soils with additives improves the engineering behavior and sustainability of soil, and reduces long-term problems including swelling and deformation.

The aim of this paper is to model the geotechnical properties of lime-stabilized fine grained-soils, as reported by Ismeik and Shaqour (2018), with preliminary results derived from an ongoing research project.

© Springer Nature Switzerland AG 2020
L. Hoyos and H. Shehata (Eds.): GeoMEast 2019, SUCI, pp. 150–157, 2020.
https://doi.org/10.1007/978-3-030-34206-7_12

2   Materials and Experimental Work

The clay used in this study is a mixture of kaolinite and montmorillonite with other minerals including muscovite, goethite, gypsum, quartz, feldspar, illite, and hematite. In accordance with AASHTO M145-91 (2012) and ASTM D 2487-11 (2011) standards, the soil is classified as A-7-5 (42) and CH, respectively. A commercial lime is used to stabilize the soil. Physical and index properties of the clay and lime used in this investigation are given in Tables 1 and 2, respectively.

Table 1. Physical and index properties of fine-grained soil

Property	Value
Color	Red
Specific gravity	2.77
Atterberg limits	
Liquid limit (%)	73
Plastic limit (%)	41
Plasticity index (%)	32
Grading	
Coefficient of uniformity, C_u	10.2
Coefficient of curvature, C_c	1.45
Effective diameter, D_{10} (mm)	0.002
Nominal mean size, D_{50} (mm)	0.018
Maximum dry density (kN/m^3)	24.21
Optimum moisture content (%)	14.70
Unconfined compressive strength (kPa)	275
Class classification	
USCS	CH
AASHTO	A-7-5

Table 2. Physical properties of lime

Property	Value
Color	White
Specific gravity	2
Over 90 μm (%)	<9
Over 630 μm (%)	0
Insoluble material (%)	<1
Bulk density (g/l)	600–900
LOI (%)	23.9
pH	12.6
Reactivity (min)	<9

The following procedure is used to prepare the samples. After drying the soil at a temperature of 80 °C, the soil is mixed thoroughly with 2, 4, 6, and 8% lime. Water is added gradually and mixed with the lime-soil mixture until paste becomes homogeneous. Samples are prepared within the proctor mold at optimum moisture content and maximum dry density.

Un-stabilized and stabilized soil samples are experimentally tested to determine index properties of soil. Namely, Atterberg limits, and density-moisture relationships tests are conducted in accordance with ASTM D4318-10 (2010) and ASTM D698-12 (2012) standards, respectively. As for the unconfined compressive strength test, cylindrical samples of 70 mm in length and 35 mm in diameter are tested in accordance with ASTM D2166-13 (2013) standard with a loading rate of 0.1 mm/min.

3   Analysis of Results

3.1   Atterberg Limits

Consistency limits results of the untreated and lime-treated soils are presented in Fig. 1. The liquid limit, plastic limit, and plasticity index for the untreated soil were found to be 66.59, 31.65, and 34.94%, respectively. Treated soil showed relatively marginal change of liquid limit with increasing lime content (1.41%). Addition of lime increased the plastic limit by 56.87% at 2% lime content and thereafter remained almost constant with further increase of lime content. Plasticity index decreased about 64.08% when soil was treated with 2% lime with no further change with incremental addition of lime.

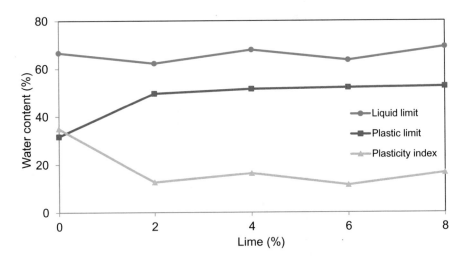

Fig. 1. Influence of lime content on Atterberg limits

Plasticity index reduction is mostly attributed to the increase of plastic limit associated with the addition of lime. Improvement levels are evident due to addition of lime. A concentration of between 2 and 8% is found to reduce the plasticity of soil. Similar results of treated soil are reported by Harichane et al. (2012), Sivapullaiah et al. (2000), and Prusinski and Bhattacharja (1999). The reduction of plasticity of the treated soil indicates that lime can be used to enhance clay workability, which expedites subsequent manipulation and placement of stabilized soil for roads and infrastructures construction.

3.2 Compaction Characteristics

The results of Proctor compaction tests, with various lime contents, are presented in Fig. 2 and summarized in Fig. 3. At 2, 4, 6, and 8% lime contents, compaction curves showed that the maximum dry density of treated soils, 12.81, 12.36, 11.90, and 12.20 kN/m^3, were lower than that of untreated soil 15.01 kN/m^3. Maximum dry density of treated soil was decreased by 14.65% with the addition of 2% lime and slightly reduced with the further addition of lime. The decrease of maximum dry density of soil was 18.72% at 8% lime content. Reduction of maximum dry density is linked to the lower value of specific gravity of lime (2) than that of soil (2.77).

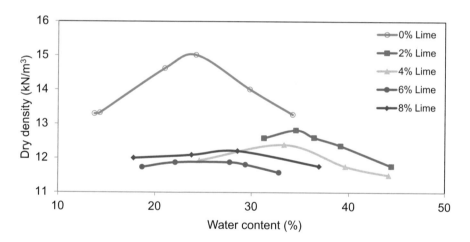

Fig. 2. Density-moisture relationships for the soil before and after lime treatment

Optimum moisture content was 42.56% higher for the stabilized soil than that of unstabilized soil at 2% lime content. Overall, that the addition of lime increased the optimum moisture content of treated soil at lime contents between 2 and 4%. Thus, as lime contents increases, optimum moisture content increases and maximum dry density decreases, which is in agreement with the studies of Jafari and Esna-ashari (2012), Harichane et al. (2012), and Al-Kiki et al. (2011).

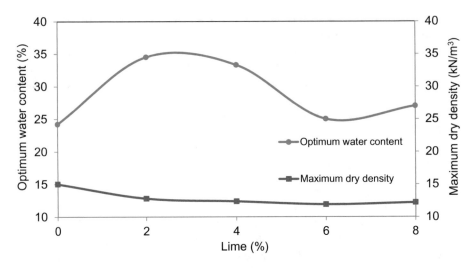

Fig. 3. Influence of lime content on compaction properties

3.3   Unconfined Compressive Strength

Strength results of the untreated and lime-treated soil samples, cured for 28 days, are shown in Fig. 4. The unconfined compressive strength values of untreated soil and 2 and 4% lime contents were 275, 452, and 606 kPa, respectively. The strength of treated samples increased with the increase of lime content until it reaches a peak value of 1632 kPa at 6% lime content. However, further addition of lime to 8% reduced the strength to 1386 kPa.

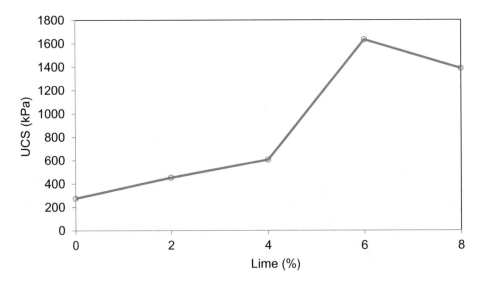

Fig. 4. Influence of lime content on the unconfined compressive strength of soil

The results show that 6% of lime achieves the maximum strength for the clay (5.93 fold). As noticed, the 8% lime strength value was slightly lower than that of 6% lime. This is explained by the negative effect of excess lime, which leads to incomplete hydration reaction between the soil and available water in the mixture. This behavior is similar to other studies reported by Jafari and Esna-ashari (2012), Lin et al. (2007), and Okagbue and Yakubu (2000).

3.4 Statistical Analyses

An attempt is made to determine the relationship between soil index properties and amount of added lime. Statistical regression techniques were employed to predict such relationships. In this context, linear and nonlinear models were developed to establish the association between the independent variable, lime content (L), and depended variables, soil properties, namely, liquid limit (LL), plasticity index (PI), optimum moisture content (OMC), maximum dry density (MDD), and unconfined compressive strength (UCS). As a result, 5 models were developed and grouped into three classes as shown below.

Atterberg limits

$$LL = 66.473 + 1.904L - 4.747L^{0.5} \quad \left(R^2 = 0.392\right) \tag{1}$$

$$PI = 34.711 + 5.441L - 21.938L^{0.5} \quad \left(R^2 = 0.937\right) \tag{2}$$

Compaction properties

$$OMC = 0.227L^3 - 3.148L^2 + 11.002L + 24.014 \quad \left(R^2 = 0.973\right) \tag{3}$$

$$MDD = 15.021 + 0.389L - 2.139L^{0.5} \quad \left(R^2 = 0.989\right) \tag{4}$$

Unconfined compressive strength

$$UCS = 297.583 + 666.141L - 516.489L^2 + 127.002L^3 - 8.833L^4 - 18.959N \quad \left(R^2 = 0.990\right) \tag{5}$$

The reliability of the models is verified with the coefficient of determination (R^2) test. As calculated, the values of R^2 for Eqs. 1 to 5 were 0.392, 0.937, 0.973, 0.989, and 0.990. Since these values are very close to 1, with exception of Eq. 1, we can conclude that these models are valid and useful to estimate reliability and efficiently the physical properties of lime-treated soils used in this study.

4   Conclusions

In this experimental investigation, an effort was carried out to improve the physical properties of fine-grained subgrades used for pavement construction. Benefits of lime treatment of fine-grained soils were quantified. As a result, plastic properties were enhanced, optimum moisture content was increased, and maximum dry density was decreased. An amount of about 6 to 8% lime content was found to greatly improve the unconfined strength of soil, which is an essential parameter for pavement design. Such treatment can significantly reduce permanent deformation (rutting) in flexible pavement due to weak subgrade soil. The suggested models were found useful to determine geotechnical properties of lime-treated clays instead of laboratory testing.

5   Future Additional Research

The work and initial results presented in this publication are preliminary in the context of an ongoing larger research project. It is early to generalize these findings for different clays since the mineralogy of soil has an effect on the treatment process. The experimental laboratory results, modeled in this publication, are obtained from the findings of Ismeik and Shaqour (2018). Additional work is underway to further investigate the durability and long-term performance of lime-treated clays and the outcomes will be published in due course.

References

AASHTO (American Association of State and Highway Transportation Officials): Standard specification for classification of soils and soil-aggregate mixtures for highway construction purposes. AASHTO M145-91, Washington, DC, USA (2012)

Al-Kiki, I., Al-Attalla, M., Al-Zubaydi, A.: Long term strength and durability of clayey soil stabilized with lime. Eng. Technol. J. **29**(4), 725–735 (2011)

ASTM (American Society for Testing and Materials): Standard test methods for liquid limit, plastic limit, and plasticity index of soils. ASTM D4318-10, West Conshohocken, PA, USA (2010)

ASTM (American Society for Testing and Materials): Standard practice for classification of soils for engineering purposes (Unified Soil Classification System). ASTM D 2487-11, West Conshohocken, PA, USA (2011)

ASTM (American Society for Testing and Materials): Fundamental principles of soil compaction. ASTM D698-12, West Conshohocken, PA, USA (2012)

ASTM (American Society for Testing and Materials): Standard test method for unconfined compressive strength of cohesive soil. ASTM D2166-13, West Conshohocken, PA, USA (2013)

Harichane, K., Ghrici, M., Kenai, S.: Effect of the combination of lime and natural pozzolana on the compaction and strength of soft clayey soils: a preliminary study. Environ. Earth Sci. **66**(8), 2197–2205 (2012)

Ismeik, M., Shaqour, F.: Effectiveness of lime in stabilising subgrade soils subjected to freeze-thaw cycles. Road Mater. Pavement Des. (2018). https://doi.org/10.1080/14680629.2018.1479289

Jafari, M., Esna-ashari, M.: Effect of waste tire cord reinforcement on unconfined compressive strength of lime stabilized clayey soil under freeze-thaw condition. Cold Reg. Sci. Technol. **82**, 21–29 (2012)

Lin, D.F., Lin, K.L., Hung, M.J., Luo, H.L.: Sludge ash/hydrated lime on the geotechnical properties of soft soil. J. Hazard. Mater. **145**(1), 58–64 (2007)

Okagbue, C.O., Yakubu, J.A.: Limestone ash waste as a substitute for lime in soil improvement for engineering construction. Bull. Eng. Geol. Env. **58**(2), 107–113 (2000)

Prusinski, J., Bhattacharja, S.: Effectiveness of Portland cement and lime in stabilizing clay soils. Transp. Res. Rec. J. Transp. Res. Board **1652**, 215–227 (1999)

Sivapullaiah, P.V., Sridharan, A., Bhaskar Raju, K.V.: Role of amount and type of clay in the lime stabilization of soils. Proc. Inst. Civ. Eng. Ground Improv. **4**(1), 37–45 (2000)

Author Index

Printed in the United States
By Bookmasters